THE
APPLIED
DATA SCIENCE
WORKSHOP

SECOND EDITION

Get started with the applications
of data science and techniques
to explore and assess data effectively

Alex Galea

THE APPLIED DATA SCIENCE WORKSHOP
SECOND EDITION

Author: Alex Galea

Reviewers: Paul Van Branteghem, Guillermina Bea Fernández, Shovon Sengupta, and Karen Yang

Managing Editor: Anushree Arun Tendulkar

Acquisitions Editors: Royluis Rodrigues and Karan Wadekar

Production Editor: Roshan Kawale

Editorial Board: Megan Carlisle, Samuel Christa, Mahesh Dhyani, Heather Gopsill, Manasa Kumar, Alex Mazonowicz, Monesh Mirpuri, Bridget Neale, Dominic Pereira, Shiny Poojary, Abhishek Rane, Brendan Rodrigues, Erol Staveley, Ankita Thakur, Nitesh Thakur, and Jonathan Wray

First published: October 2018

Second edition: July 2020

Production reference: 2230221

ISBN: 978-1-80020-250-4

Published by Packt Publishing Ltd.
Livery Place, 35 Livery Street
Birmingham B3 2PB, UK

WHY LEARN WITH A PACKT WORKSHOP?

LEARN BY DOING

Packt Workshops are built around the idea that the best way to learn something new is by getting hands-on experience. We know that learning a language or technology isn't just an academic pursuit. It's a journey towards the effective use of a new tool—whether that's to kickstart your career, automate repetitive tasks, or just build some cool stuff.

That's why Workshops are designed to get you writing code from the very beginning. You'll start fairly small—learning how to implement some basic functionality—but once you've completed that, you'll have the confidence and understanding to move onto something slightly more advanced.

As you work through each chapter, you'll build your understanding in a coherent, logical way, adding new skills to your toolkit and working on increasingly complex and challenging problems.

CONTEXT IS KEY

All new concepts are introduced in the context of realistic use-cases, and then demonstrated practically with guided exercises. At the end of each chapter, you'll find an activity that challenges you to draw together what you've learned and apply your new skills to solve a problem or build something new.

We believe this is the most effective way of building your understanding and confidence. Experiencing real applications of the code will help you get used to the syntax and see how the tools and techniques are applied in real projects.

BUILD REAL-WORLD UNDERSTANDING

Of course, you do need some theory. But unlike many tutorials, which force you to wade through pages and pages of dry technical explanations and assume too much prior knowledge, Workshops only tell you what you actually need to know to be able to get started making things. Explanations are clear, simple, and to-the-point. So you don't need to worry about how everything works under the hood; you can just get on and use it.

Written by industry professionals, you'll see how concepts are relevant to real-world work, helping to get you beyond "Hello, world!" and build relevant, productive skills. Whether you're studying web development, data science, or a core programming language, you'll start to think like a problem solver and build your understanding and confidence through contextual, targeted practice.

ENJOY THE JOURNEY

Learning something new is a journey from where you are now to where you want to be, and this Workshop is just a vehicle to get you there. We hope that you find it to be a productive and enjoyable learning experience.

Packt has a wide range of different Workshops available, covering the following topic areas:

- Programming languages

- Web development

- Data science, machine learning, and artificial intelligence

- Containers

Once you've worked your way through this Workshop, why not continue your journey with another? You can find the full range online at http://packt.live/2MNkuyl.

If you could leave us a review while you're there, that would be great. We value all feedback. It helps us to continually improve and make better books for our readers, and also helps prospective customers make an informed decision about their purchase.

Thank you,
The Packt Workshop Team

Table of Contents

Chapter 3: Preparing Data for Predictive Modeling   87

Chapter 4: Training Classification Models   125

PREFACE

ABOUT THE BOOK

From banking and manufacturing through to education and entertainment, using data science for business has revolutionized almost every sector in the modern world. It has an important role to play in everything from app development to network security.

Taking an interactive approach to learning the fundamentals, this book is ideal for beginners. You'll learn all the best practices and techniques for applying data science in the context of real-world scenarios and examples.

Starting with an introduction to data science and machine learning, you'll start by getting to grips with Jupyter functionality and features. You'll use Python libraries like scikit-learn, pandas, Matplotlib, and Seaborn to perform data analysis and data preprocessing on real-world datasets from within your own Jupyter environment. Progressing through the chapters, you'll train classification models using scikit-learn, and assess model performance using advanced validation techniques. Towards the end, you'll use Jupyter Notebooks to document your research, build stakeholder reports, and even analyze web performance data.

By the end of *The Applied Data Science Workshop*, *Second Edition*, you'll be prepared to progress from being a beginner to taking your skills to the next level by confidently applying data science techniques and tools to real-world projects.

AUDIENCE

If you are an aspiring data scientist who wants to build a career in data science or a developer who wants to explore the applications of data science from scratch and analyze data in Jupyter using Python libraries, then this book is for you. Although a brief understanding of Python programming and machine learning is recommended to help you grasp the topics covered in the book more quickly, it is not mandatory.

ABOUT THE CHAPTERS

Chapter 1, Introduction to Jupyter Notebooks, will get you started by explaining how to use the Jupyter Notebook and JupyterLab platforms. After going over the basics, we will discuss some fantastic features of Jupyter, which include tab completion, magic functions, and new additions to the JupyterLab interface. Finally, we will look at the Python libraries we'll be using in this book, such as pandas, seaborn, and scikit-learn.

Chapter 2, Data Exploration with Jupyter, is focused on exploratory analysis in a live Jupyter Notebook environment. Here, you will use visualizations such as scatter plots, histograms, and violin plots to deepen your understanding of the data. We will also walk through some simple modeling problems with scikit-learn.

Chapter 3, Preparing Data for Predictive Modeling, will enable you to plan a machine learning strategy and assess whether or not data is suitable for modeling. In addition to this, you'll learn about the process involved in preparing data for machine learning algorithms, and apply this process to sample datasets using pandas.

Chapter 4, Training Classification Models, will introduce classification algorithms such as SVMs, KNNs, and Random Forests. Using a real-world Human Resources analytics dataset, we'll train and compare models that predict whether an employee will leave their company. You'll learn about training models with scikit-learn and use decision boundary plots to see what overfitting looks like.

Chapter 5, Model Validation and Optimization, will give you hands-on experience with model testing and model selection concepts, including k-fold cross-validation and validation curves. Using these techniques, you'll learn how to optimize model parameters and compare model performance reliably. You will also learn how to implement dimensionality reduction techniques such as Principal Component Analysis (PCA).

Chapter 6, Web Scraping with Jupyter Notebooks, will focus on data acquisition from online sources such as web pages and APIs. You will see how data can be downloaded from the web using HTTP requests and HTML parsing. After collecting data in this way, you'll also revisit concepts learned in earlier chapters, such as data processing, analysis, visualization, and modeling.

CONVENTIONS

Code words in text, database table names, folder names, filenames, file extensions, pathnames, dummy URLs, user input, and Twitter handles are shown as follows:

"It's recommended to install some of these (such as `mlxtend`, `watermark`, and `graphviz`) ahead of time if you have access to an internet connection now. This can be done by opening a new Terminal window and running the `pip` or `conda` commands."

Words that you see on the screen (for example, in menus or dialog boxes) appear in the same format.

A block of code is set as follows:

```
https://github.com/rasbt/mlxtend
pip install mlxtend
```

New terms and important words are shown like this:

"The focus of this chapter is to introduce **Jupyter Notebooks**—the data science tool that we will be using throughout the book."

CODE PRESENTATION

Lines of code that span multiple lines are split using a backslash (\ ). When the code is executed, Python will ignore the backslash, and treat the code on the next line as a direct continuation of the current line.

For example:

```
history = model.fit(X, y, epochs=100, batch_size=5, verbose=1, \
                    validation_split=0.2, shuffle=False)
```

Comments are added into code to help explain specific bits of logic. Single-line comments are denoted using the # symbol, as follows:

```
# Print the sizes of the dataset
print("Number of Examples in the Dataset = ", X.shape[0])
print("Number of Features for each example = ", X.shape[1])
```

Multi-line comments are enclosed by triple quotes, as shown below:

```
"""
Define a seed for the random number generator to ensure the
result will be reproducible
"""
seed = 1
np.random.seed(seed)
random.set_seed(seed)
```

SETTING UP YOUR ENVIRONMENT

Before we explore the book in detail, we need to set up specific software and tools. In the following section, we shall see how to do that.

INSTALLING PYTHON

The easiest way to get up and running with this workshop is to install the Anaconda Python distribution. This can be done as follows:

1. Navigate to the Anaconda downloads page from https://www.anaconda.com/.

2. Download the most recent Python 3 distribution for your operating system – currently, the most stable version is Python 3.7.

3. Open and run the installation package. If prompted, select **yes** for the option to **Register Anaconda as my default Python**.

INSTALLING LIBRARIES

pip comes pre-installed with Anaconda. Once Anaconda is installed on your machine, all the required libraries can be installed using **pip**, for example, **pip install numpy**. Alternatively, you can install all the required libraries using **pip install -r requirements.txt**. You can find the **requirements.txt** file at https://packt.live/2YBPK5y.

The exercises and activities will be executed in Jupyter Notebooks. Jupyter is a Python library and can be installed in the same way as the other Python libraries – that is, with **pip install jupyter**, but fortunately, it comes pre-installed with Anaconda. To open a notebook, simply run the command **jupyter notebook** in the Terminal or Command Prompt.

WORKING WITH JUPYTERLAB AND JUPYTER NOTEBOOK

You'll be working on different exercises and activities using either the JupyterLab or Jupyter Notebook platforms. These exercises and activities can be downloaded from the associated GitHub repository.

Download the repository from https://packt.live/2zwhfom.

You can either clone it using **git** or download it as a zipped folder by clicking on the green **Clone or download** button in the upper-right corner.

In order to launch a Jupyter Notebook workbook, you should first use the Terminal to navigate to your source code. See the following, for example:

```
cd The-Applied-Data-Science-Workshop
```

Once you are in the project directory, simply run **jupyter lab** to start up JupyterLab. Similarly, for Jupyter Notebook, run **jupyter notebook**.

ACCESSING THE CODE FILES

You can find the complete code files of this book at https://packt.live/2zwhfom. You can also run many activities and exercises directly in your web browser by using the interactive lab environment at https://packt.live/3d6yr1A.

We've tried to support interactive versions of all activities and exercises, but we recommend a local installation as well for instances where this support isn't available.

If you have any issues or questions about installation, please email us at workshops@packt.com.

1

INTRODUCTION TO JUPYTER NOTEBOOKS

OVERVIEW

This chapter describes Jupyter Notebooks and their use in data analysis. It also explains the features of Jupyter Notebooks, which allow for additional functionality beyond running Python code. You will learn and implement the fundamental features of Jupyter Notebooks by completing several hands-on exercises. By the end of this chapter, you will be able to use some important features of Jupyter Notebooks and some key libraries available in Python.

INTRODUCTION

Our approach to learning in this book is highly applied since hands-on learning is the quickest way to understand abstract concepts. With this in mind, the focus of this chapter is to introduce **Jupyter Notebooks**—the data science tool that we will be using throughout this book.

Since Jupyter Notebooks have gained mainstream popularity, they have been one of the most important tools for data scientists who use Python. This is because they offer a great environment for a variety of tasks, such as performing quick and dirty analysis, researching model selection, and creating reproducible pipelines. They allow for data to be loaded, transformed, and modeled inside a single file, where it's quick and easy to test out code and explore ideas along the way. Furthermore, all of this can be documented **inline** using formatted text, which means you can make notes or even produce a structured report.

Other comparable platforms—for example, **RStudio** or **Spyder**—offer multiple panels to work between. Frequently, one of these panels will be a **Read Eval Prompt Loop (REPL)**, where code is run on a Terminal session that has saved memory. Code written here may end up being copied and pasted into a different panel within the main codebase, and there may also be additional panels to see visualizations or other files. Such development environments are prone to efficiency issues and can promote bad practices for reproducibility if you're not careful.

Jupyter Notebooks work differently. Instead of having multiple panels for different components of your project, they offer the same functionality in a single component (that is, the Notebook), where the text is displayed along with code snippets, and code outputs are displayed inline. This lets you code efficiently and allows you to look back at previous work for reference, or even make alterations.

We'll start this chapter by explaining exactly what Jupyter Notebooks are and why they are so popular among data scientists. Then, we'll access a Notebook together and go through some exercises to learn how the platform is used.

BASIC FUNCTIONALITY AND FEATURES OF JUPYTER NOTEBOOKS

In this section, we will briefly demonstrate the usefulness of Jupyter Notebooks with examples. Then, we'll walk through the basics of how they work and how to run them within the Jupyter platform. For those who have used Jupyter Notebooks before, this will be a good refresher, and you are likely to uncover new things as well.

WHAT IS A JUPYTER NOTEBOOK AND WHY IS IT USEFUL?

Jupyter Notebooks are locally run on web applications that contain live code, equations, figures, interactive apps, and Markdown text in which the default programming language is Python. In other words, a Notebook will assume you are writing Python unless you tell it otherwise. We'll see examples of this when we work through our first workbook, later in this chapter.

> **NOTE**
>
> Jupyter Notebooks support many programming languages through the use of kernels, which act as bridges between the Notebook and the language. These include R, C++, and JavaScript, among many others. A list of available kernels can be found here: https://packt.live/2Y0jKJ0.

The following is an example of a Jupyter Notebook:

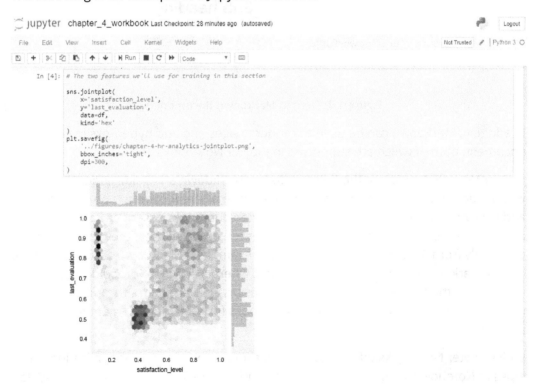

Figure 1.1: Jupyter Notebook sample workbook

Besides executing Python code, you can write in Markdown to quickly render formatted text, such as titles, lists, or bold font. This can be done in combination with code using the concept of independent cells in the Notebook, as seen in *Figure 1.2*. Markdown is not specific to Jupyter; it is also a simple language used for styling text and creating basic documents. For example, most GitHub repositories have a **README.md** file that is written in Markdown format. It's comparable to HTML but offers much less customization in exchange for simplicity.

Commonly used symbols in markdown include hashes (**#**) to make text into a heading, square (**[]**) and round brackets (**()**) to insert hyperlinks, and asterisks (*****) to create italicized or bold text:

Figure 1.2: Sample Markdown document

In addition, Markdown can be used to render images and add hyperlinks in your document, both of which are supported in Jupyter Notebooks.

Jupyter Notebooks was not the first tool to use Markdown alongside code. This was the design of **R Markdown**, a hybrid language where R code can be written and executed inline with Markdown text. Jupyter Notebooks essentially offer the equivalent functionality for Python code. However, as we will see, they function quite differently from R Markdown documents. For example, R Markdown assumes you are writing Markdown unless otherwise specified, whereas Jupyter Notebooks assume you are inputting code. This and other features (as we will explore throughout) make it more appealing to use Jupyter Notebooks for rapid development in data science research.

While Jupyter Notebooks offer a blank canvas for a general range of applications, the types of Notebooks commonly seen in real-world data science can be categorized as either lab-style or deliverable.

Lab-style Notebooks serve as the programming analog of research journals. These should contain all the work you've done to load, process, analyze, and model the data. The idea here is to document everything you've done for future reference. For this reason, it's usually not advisable to delete or alter previous lab-style Notebooks. It's also a good idea to accumulate multiple date-stamped versions of the Notebook as you progress through the analysis, in case you want to look back at previous states.

Deliverable Notebooks are intended to be presentable and should contain only select parts of the lab-style Notebooks. For example, this could be an interesting discovery to share with your colleagues, an in-depth report of your analysis for a manager, or a summary of the key findings for stakeholders.

In either case, an important concept is **reproducibility**. As long as all the relevant software versions were documented at runtime, anybody receiving a Notebook can rerun it and compute the same results as before. The process of actually running code in a Notebook (as opposed to reading a pre-computed version) brings you much closer to the actual data. For example, you can add cells and ask your own questions regarding the datasets or tweak existing code. You can also experiment with Python to break down and learn about sections of code that you are struggling to understand.

EDITING NOTEBOOKS WITH JUPYTER NOTEBOOKS AND JUPYTERLAB

It's finally time for our first exercise. We'll start by exploring the interface of the Jupyter Notebook and the JupyterLab platforms. These are very similar applications for running Jupyter Notebook (`.ipynb`) files, and you can use whatever platform you prefer for the remainder of this book, or swap back and forth, once you've finished the following exercises.

> **NOTE**
>
> The `.ipynb` file extension is standard for Jupyter Notebooks, which was introduced back when they were called IPython Notebooks. These files are human-readable JSON documents that can be opened and modified with any text editor. However, there is usually no reason to open them with any software other than Juptyer Notebook or JupyterLab, as described in this section. Perhaps the one exception to this rule is when doing version control with Git, if you may want to see the changes in plain text.

At this stage, you'll need to make sure that you have the companion material downloaded. This can be downloaded from the open source repository on GitHub at https://packt.live/2zwhfom.

In order to run the code, you should download and install the Anaconda Python distribution for Python 3.7 (or a more recent version). If you already have Python installed and don't want to use Anaconda, you may choose to install the dependencies manually instead (see **requirements.txt** in the GitHub repository).

> **NOTE**
>
> Virtual environments are a great tool for managing multiple projects on the same machine. Each virtual environment may contain a different version of Python and external libraries. In addition to Python's built-in virtual environments, **conda** also offers virtual environments, which tend to integrate better with Jupyter Notebooks.
>
> For the purposes of this book, you do not need to worry about virtual environments. This is because they add complexity that will likely lead to more issues than they aim to solve. Beginners are advised to run global system installs of Python libraries (that is, using the **pip** commands shown here). However, more experienced Python programmers might wish to create and activate a virtual environment for this project.

We will install additional Python libraries throughout this book, but it's recommended to install some of these (such as **mlxtend**, **watermark**, and **graphviz**) ahead of time if you have access to an internet connection now. This can be done by opening a new Terminal window and running the **pip** or **conda** commands, as follows:

- **mlxtend** (https://packt.live/3ftcN98): This is a useful tool for particular data science tasks. We'll use it to visualize the decision boundaries of models in *Chapter 5, Model Validation and Optimization*, and *Chapter 6, Web Scraping with Jupyter Notebooks*:

```
pip install mlxtend
```

- **watermark** (https://packt.live/2N1qjok): This IPython magic extension is used for printing version information. We'll use it later in this chapter:

```
pip install watermark
```

- **graphviz** (https://packt.live/3hqqCHz): This is for rendering graph visualizations. We'll use this for visualizing decision trees in *Chapter 5, Model Validation and Optimization*:

```
conda install -c anaconda graphviz python-graphviz
```

graphviz will only be used once, so don't worry too much if you have issues installing it. However, hopefully, you were able to get **mlxtend** installed since we'll need to rely on it in later chapters to compare models and visualize how they learn patterns in the data.

EXERCISE 1.01: INTRODUCING JUPYTER NOTEBOOKS

In this exercise, we'll launch the Jupyter Notebook platform from the Terminal and learn how the visual user interface works. Follow these steps to complete this exercise:

1. Navigate to the companion material directory from the Terminal. If you don't have the code downloaded yet, you can clone it using the **git** command-line tool:

```
git clone https://github.com/PacktWorkshops/The-Applied-Data-Science-
Workshop.git
cd The-Applied-Data-Science-Workshop
```

> **NOTE**
>
> With Unix machines such as Mac or Linux, command-line navigation can be done using **ls** to display directory contents and **cd** to change directories. On Windows machines, use **dir** to display directory contents and use **cd** to change directories. If, for example, you want to change the drive from **C:** to **D:**, you should execute **D:** to change drives. This is an important step if you wish to enable all commands based on folder structure and ensure they run smoothly.

2. Run the Jupyter Notebook platform by asking for its version:

> **NOTE**
>
> The # symbol in the code snippet below denotes a code comment.
> Comments are added into code to help explain specific bits of logic.

```
jupyter notebook --version
# should return 6.0.2 or a similar / more recent version
```

3. Start a new local Notebook server here by typing the following into the Terminal:

```
jupyter notebook
```

A new window or tab of your default browser will open the Notebook Dashboard to the working directory. Here, you will see a list of folders and files contained therein.

4. Reopen the Terminal window that you used to launch the app. We will see the NotebookApp being run on a local server. In particular, you should see a line like this in the Terminal:

```
[I 20:03:01.045 NotebookApp] The Jupyter Notebook is running at:
http:// localhost:8888/?token=e915bb06866f19ce462d959a9193a94c7
c088e81765f9d8a
```

Going to the highlighted HTTP address will load the app in your browser window, as was done automatically when starting the app.

5. Reopen the web browser and play around with the Jupyter Dashboard platform by clicking on a folder (such as **chapter-06**), and then clicking on an **.ipynb** file (such as **chapter_6_workbook.ipynb**) to open it. This will cause the Notebook to open in a new tab on your browser.

6. Go back to the tab on your browser that contains the Jupyter Dashboard. Then, go back to the root directory by clicking the ... button (above the folder content listing) or the folder icon above that (in the current directory breadcrumb).

7. Although its main use is for editing Notebook files, Jupyter is a basic text editor as well. To see this, click on the **requirements.txt** text file. Similar to the Notebook file, it will open in a new tab of your browser.

8. Now, you need to close the platform. Reopen the Terminal window you used to launch the app and stop the process by typing *Ctrl + C* in the Terminal. You may also have to confirm this by entering **y** and pressing *Enter*. After doing this, close the web browser window as well.

9. Now, you are going to explore the Jupyter Notebook **command-line interface (CLI)** a bit. Load the list of available options by running the following command:

```
jupyter notebook --help
```

10. One option is to specify the port for the application to run on. Open the NotebookApp at local port **9000** by running the following command:

```
jupyter notebook --port 9000
```

11. Click **New** in the upper right-hand corner of the Jupyter Dashboard and select a kernel from the drop-down menu (that is, select something in the **Notebooks** section):

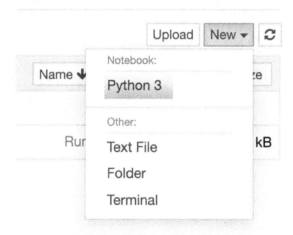

Figure 1.3: Selecting a kernel from the drop-down menu

This is the primary method of creating a new Jupyter Notebook.

Kernels provide programming language support for the Notebook. If you have installed Python with Anaconda, that version should be the **default** kernel. Virtual environments that have been properly configured will also be available here.

12. With the newly created blank Notebook, click the top cell and type
    `print('hello world')`, or any other code snippet that writes to the screen.

13. Click the cell and press *Shift + Enter* or select **Run Cell** from the **Cell** menu.

    Any **stdout** or **stderr** output from the code will be displayed beneath as the
    cell runs. Furthermore, the string representation of the object written in the final
    line will be displayed as well. This is very handy, especially for displaying tables,
    but sometimes, we don't want the final object to be displayed. In such cases, a
    semicolon (;) can be added to the end of the line to suppress the display. New
    cells expect and run code input by default; however, they can be changed to
    render markdown instead.

14. Click an empty cell and change it to accept the Markdown-formatted text. This
    can be done from the drop-down menu icon in the toolbar or by selecting
    Markdown from the **Cell** menu. Write some text in here (any text will do),
    making sure to utilize Markdown formatting symbols such as **#**, and then run the
    cell using *Shift + Enter*:

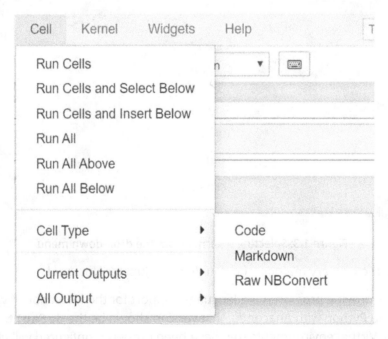

Figure 1.4: Menu options for converting cells into code/Markdown

15. Scroll to the **Run** button in the toolbar:

Figure 1.5: Toolbar icon to start cell execution

16. This can be used to run cells. As you will see later, however, it's handier to *Shift + Enter* to run cells.

17. Right next to the Run button is a **Stop** icon, which can be used to stop cells from running. This is useful, for example, if a cell is taking too long to run:

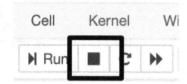

Figure 1.6: Toolbar icon to stop cell execution

18. New cells can be manually added from the **Insert** menu:

Figure 1.7: Menu options for adding new cells

19. Cells can be copied, pasted, and deleted using icons or by selecting options from the **Edit** menu:

Figure 1.8: Toolbar icons to cut, copy, and paste cells

The drop-down list from the **Edit** menu is as follows:

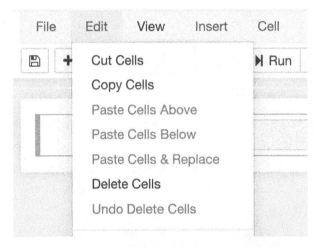

Figure 1.9: Menu options to cut, copy, and paste cells

20. Cells can also be moved up and down this way:

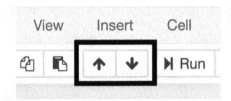

Figure 1.10: Toolbar icons for moving cells up or down

There are useful options in the **Cell** menu that you can use to run a group of cells or the entire Notebook:

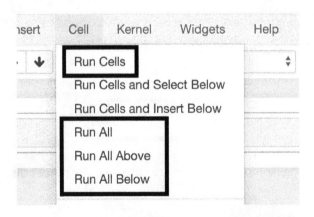

Figure 1.11: Menu options for running cells in bulk

Experiment with the toolbar options to move cells up and down, insert new cells, and delete cells. An important thing to understand about these Notebooks is the shared memory between cells. It's quite simple; every cell that exists on the sheet has access to the global set of variables. So, for example, a function defined in one cell could be called from any other, and the same applies to variables. As you would expect, anything within the scope of a function will not be a global variable and can only be accessed from within that specific function.

21. Open the **Kernel** menu to see the selections. The **Kernel** menu is useful for stopping the execution of the script and restarting the Notebook if the kernel dies:

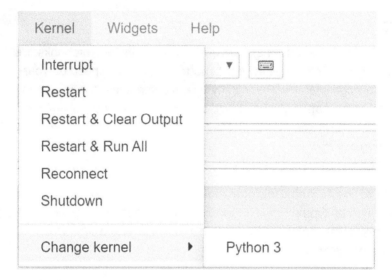

Figure 1.12: Menu options for selecting a Notebook kernel

Kernels can also be swapped here at any time, but it is inadvisable to use multiple kernels for a single Notebook due to reproducibility concerns.

22. Open the **File** menu to see the selections. The **File** menu contains options for downloading the Notebook in various formats. It's recommended to save an HTML version of your Notebook, where the content is rendered statically and can be opened and viewed as you would expect in web browsers.

23. The Notebook name will be displayed in the upper left-hand corner. New Notebooks will automatically be named **Untitled**. You can change the name of your **.ipynb** Notebook file by clicking on the current name in the upper left corner and typing in the new name. Then, save the file.

24. Close the current tab in your web browser (exiting the Notebook) and go to the Jupyter Dashboard tab, which should still be open. If it's not open, then reload it by copying and pasting the HTTP link from the Terminal.

25. Since you didn't shut down the Notebook (you just saved and exited it), it will have a green book symbol next to its name in the **Files** section of the Jupyter Dashboard, and it will be listed as **Running** on the right-hand side next to the last modified date. Notebooks can be shut down from here.

26. Quit the Notebook you have been working on by selecting it (checkbox to the left of the name) and clicking the orange **Shutdown** button.

> **NOTE**
>
> If you plan to spend a lot of time working with Jupyter Notebooks, it's worthwhile learning the keyboard shortcuts. This will speed up your workflow considerably. Particularly useful commands to learn are the shortcuts for manually adding new cells and converting cells from code into Markdown formatting. Click on **Keyboard Shortcuts** from the **Help** menu to see how.

27. Go back to the Terminal window that's running the Jupyter Notebook server and shut it down by typing *Ctrl + C*. Confirm this operation by typing **y** and pressing *Enter*. This will automatically exit any kernel that is running. Do this now and close the browser window as well.

Now that we have learned the basics of Jupyter Notebooks, we will launch and explore the JupyterLab platform.

While the Jupyter Notebook platform is lightweight and simple by design, JupyterLab is closer to R Studio in design. In JupyterLab, you can stack notebooks side by side, along with console environments (REPLs) and data tables, among other things you may want to look at.

Although the new features it provides are nice, the simplicity of the Jupyter Notebook interface means that it's still an appealing choice. Aside from its simplicity, you may find the Jupyter Notebook platform preferable for the following reasons:

- You may notice minor latency issues in JupyterLab that are not present in the Jupyter Notebook platform.

- JupyterLab can be extremely slow to load large `.ipynb` files (this is an open issue on GitHub, as of early 2020).

Please don't let these small issues hold you back from trying out JupyterLab. In fact, it would not be surprising if you decide to use it for running the remainder of the exercises and activities in this book.

The future of open source tooling around Python and data science is going to be very exciting, and there are sure to be plenty of developments regarding Jupyter tools in the years to come. This is all thanks to the open source programmers who build and maintain these projects and the companies that contribute to the community.

EXERCISE 1.02: INTRODUCING THE JUPYTERLAB PLATFORM

In this exercise, we'll launch the JupyterLab platform and see how it compares with the Jupyter Notebook platform.

Follow these steps to complete this exercise:

1. Run JupyterLab by asking for its version:

```
jupyter lab --version
# should return 1.2.3 or a similar / more recent version
```

2. Navigate to the root directory, and then, launch JupyterLab by typing the following into the Terminal:

```
jupyter lab
```

Similar to when we ran the Jupyter Notebook server, a new window or tab on your default browser should open the JupyterLab Dashboard. Here, you will see a list of folders and files in the working directory in a navigation bar to the left:

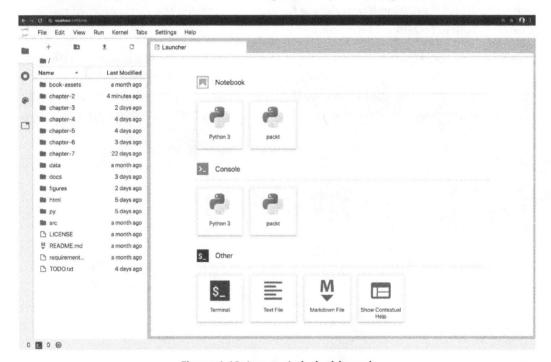

Figure 1.13: JupyterLab dashboard

3. Looking back at the Terminal, you can see a very similar output to what our NotebookApp showed us before, except now for the LabApp. If nothing else is running there, it should launch on port **8888** by default:

```
[I 18:37:29.369 LabApp] The Jupyter Notebook is running at:
[I 18:37:29.369 LabApp] http://
localhost:8888/?token=cb55c8f3c03f0d6843ae59e70bedbf3b6ec
4a92288e65fa3
```

4. Looking back at the browser window, you can see that the JupyterLab Dashboard has many of the same menus as the Jupyter Notebook platform. Open a new Notebook by clicking **File** | **New** | **Notebook**:

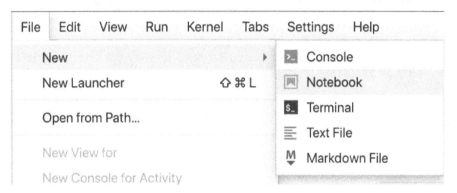

Figure 1.14: Opening a new notebook

5. When prompted to select a kernel, choose **Python 3**:

Figure 1.15: Selecting a kernel for our notebook

The Notebook will then load into a new tab inside JupyterLab. Notice how this is different from the Jupyter Notebook platform, where each file is opened in its own browser tab.

6. You will see that a toolbar has appeared at the top of the tab, with the buttons we previously explored, such as those to save, run, and stop code:

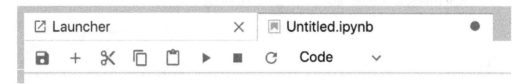

Figure 1.16: JupyterLab toolbar and Notebook tab

7. Run the following code in the first cell of the Notebook to produce some output in the space below by *Shift + Enter*:

```
for i in range(10):
    print(i, i % 3)
```

This will look as follows in Jupyter Notebook:

```
[1]: for i in range(10):
        print(i, i % 3)
```

```
0 0
1 1
2 2
3 0
4 1
5 2
6 0
7 1
8 2
9 0
```

Figure 1.17: Output of the for loop

8. When you place your mouse pointer in the white space present to the left of the cell, you will see two blue bars appear to the left of the cell. This is one of JupyterLab's new features. Click on them to hide the code cell or its output:

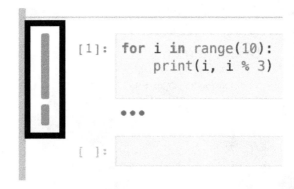

Figure 1.18: Bars that hide/show cells and output in JupyterLab

9. Explore window stacking in JupyterLab. First, save your new Notebook file by clicking **File | Save Notebook As** and giving it the name **test.ipynb**:

Figure 1.19: Prompt for saving the name of the file

10. Click **File | New | Console** in order to load up a Python interpreter session:

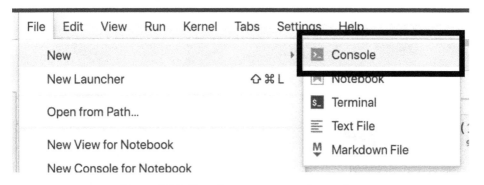

Figure 1.20: Opening a new console session

11. This time, when you see the kernel prompt, select **test.ipynb** under **Use Kernel from Other Session**. This feature of JupyterLab allows each process to have shared access to variables in memory:

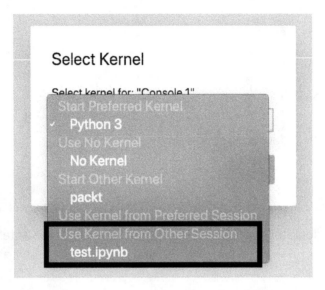

Figure 1.21: Electing the console kernel

12. Click on the new **Console** window tab and drag it down to the bottom half of the screen in order to stack it underneath the Notebook. Now, define something in the console session, such as the following:

```
a = 'apple'
```

It will look as follows:

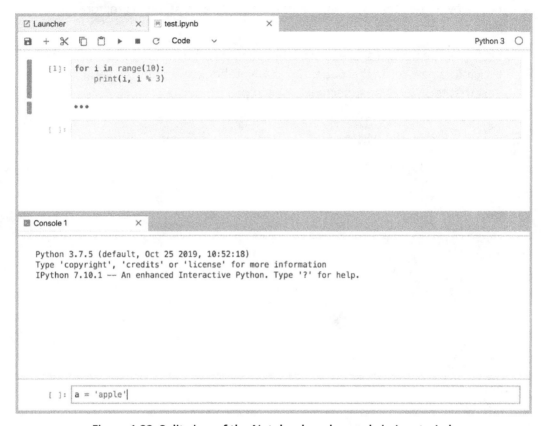

Figure 1.22: Split view of the Notebook and console in JupyterLab

13. Run this cell with *Shift + Enter* (or using the **Run** menu), and then run another cell below to test that your variable returns the value as expected; for example, **print(a)**.

14. Since you are using a shared kernel between this console and the Notebook, click into a new cell in the **test.ipynb** Notebook and print the variable there. Test that this works as expected; for example, **print(a)**:

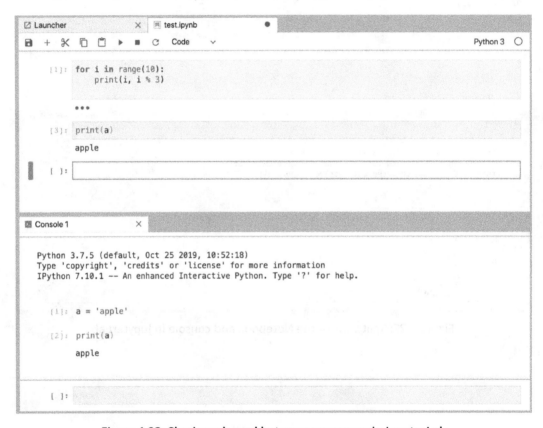

Figure 1.23: Sharing a kernel between processes in JupyterLab

A great feature of JupyterLab is that you can open up and work on multiple views of the same Notebook concurrently—something that cannot be done with the Jupyter Notebook platform. This can be very useful when working in large Notebooks where you want to frequently look at different sections.

15. You can work on multiple views of **`test.ipynb`** by right-clicking on its tab and selecting **`New View for Notebook`**:

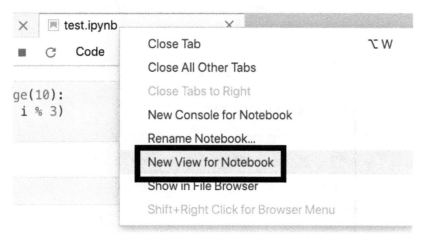

Figure 1.24: Opening a new view for an open Notebook

You should see a copy of the Notebook open to the right. Now, start typing something into one of the cells and watch the other view update as well:

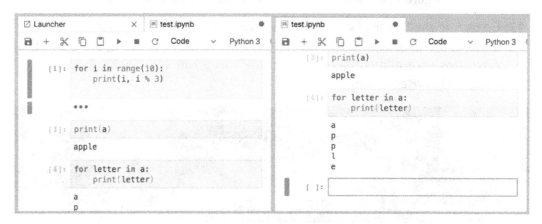

Figure 1.25: Two live views of the same Notebook in JupyterLab

There are plenty of other neat features in JupyterLab that you can discover and play with. For now, though, we are going to close down the platform.

16. Click the circular button with a box in the middle on the far left-hand side of the Dashboard. This will pull up a panel showing the kernel sessions open right now. You can click **SHUT DOWN** to close anything that is open:

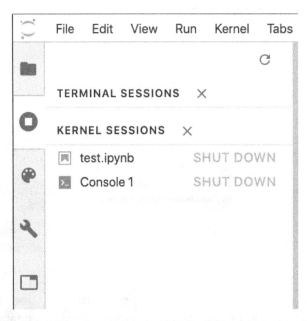

Figure 1.26: Shutting down Notebook sessions in JupyterLab

17. Go back to the Terminal window that's running the JupyterLab server and shut it down by pressing *Ctrl + C*, then confirm the operation by pressing **Y** and pressing *Enter*. This will automatically exit any kernel that is running. Do this now and close the browser window as well:

```
^C[I 11:25:39.389 LabApp] interrupted
Serving notebooks from local directory: /Users/alex/Documents/Applied-Data-Science-with-Python-and-Jupyter
1 active kernel
The Jupyter Notebook is running at:
http://localhost:8888/?token=3bbe1d5a8ce38b02050241b26cb5f5192c99db6ade530290
 or http://127.0.0.1:8888/?token=3bbe1d5a8ce38b02050241b26cb5f5192c99db6ade530290
Shutdown this notebook server (y/[n])? y
[C 11:25:40.568 LabApp] Shutdown confirmed
[I 11:25:40.569 LabApp] Shutting down 1 kernel
[I 11:25:40.775 LabApp] Kernel shutdown: 2162d7e2-4850-40f5-8744-0239076696ce
(.venv) (base) → Applied-Data-Science-with-Python-and-Jupyter git:(master) x ▮
```

Figure 1.27: Shutting down the LabApp

In this exercise, we learned about the JupyterLab platform and how it compares to the older Jupyter Notebook platform for running Notebooks. In addition to learning about the basics of using the app, we explored some of its awesome features, all of which can help your data science workflow.

In the next section, we'll learn about some of the more general features of Jupyter that apply to both platforms.

JUPYTER FEATURES

Having familiarized ourselves with the interface of two platforms for running Notebooks (Jupyter Notebook and JupyterLab), we are ready to start writing and running some more interesting examples.

> **NOTE**
>
> For the remainder of this book, you are welcome to use either the Jupyter Notebook platform or JupyterLab to follow along with the exercises and activities. The experience is similar, and you will be able to follow along seamlessly either way. Most of the screenshots for the remainder of this book have been taken from JupyterLab.

Jupyter has many appealing core features that make for efficient Python programming. These include an assortment of things, such as tab completion and viewing docstrings—both of which are very handy when writing code in Jupyter. We will explore these and more in the following exercise.

> **NOTE**
>
> The official IPython documentation can be found here: https://ipython.readthedocs.io/en/stable/. It provides details of the features we will discuss here, as well as others.

EXERCISE 1.03: DEMONSTRATING THE CORE JUPYTER FEATURES

In this exercise, we'll relaunch the Jupyter platform and walk through a Notebook to learn about some core features, such as navigating workbooks with keyboard shortcuts and using magic functions. Follow these steps to complete this exercise:

1. Start up one of the following platforms for running Jupyter Notebooks:

    JupyterLab (run **jupyter lab**)

    Jupyter Notebook (run **jupyter notebook**)

    Then, open the platform in your web browser by copying and pasting the URL, as prompted in the Terminal.

    > **NOTE**
    >
    > Here's the list of basic keyboard shortcuts; these are especially helpful if you wish to avoid having to use the mouse so often, which will greatly speed up your workflow.
    >
    > *Shift + Enter* to run a cell
    >
    > *Esc* to leave a cell
    >
    > *a* to add a cell above
    >
    > *b* to add a cell below
    >
    > *dd* to delete a cell
    >
    > *m* to change a cell to Markdown (after pressing *Esc*)
    >
    > *y* to change a cell to code (after pressing *Esc*)
    >
    > Arrow keys to move cells (after pressing *Esc*)
    >
    > *Enter* to enter a cell

    You can get help by adding a question mark to the end of any object and running the cell. Jupyter finds the docstring for that object and returns it in a pop-up window at the bottom of the app.

2. Import **numpy** and get the **arrange** docstring, as follows:

```
import numpy as np
np.arange?
```

The output will be similar to the following:

```
Docstring:
arange([start,] stop[, step,], dtype=None)

Return evenly spaced values within a given interval.

Values are generated within the half-open interval ``[start, stop)``
(in other words, the interval including `start` but excluding `stop`).
For integer arguments the function is equivalent to the Python built-in
`range` function, but returns an ndarray rather than a list.

When using a non-integer step, such as 0.1, the results will often not
be consistent.  It is better to use `numpy.linspace` for these cases.

Parameters
----------
start : number, optional
    Start of interval.  The interval includes this value.  The default
    start value is 0.
stop : number
    End of interval.  The interval does not include this value, except
    in some cases where `step` is not an integer and floating point
    round-off affects the length of `out`.
```

Figure 1.28: The docstring for np.arange

3. Get the Python **sorted** function docstring as follows:

```
sorted?
```

The output is as follows:

```
Signature: sorted(iterable, /, *, key=None, reverse=False)
Docstring:
Return a new list containing all items from the iterable in ascending order.

A custom key function can be supplied to customize the sort order, and the
reverse flag can be set to request the result in descending order.
Type:      builtin_function_or_method
```

Figure 1.29: The docstring for sort

4. You can pull up a list of the available functions on an object. You can do this for a NumPy array by running the following command:

```
a = np.array([1, 2, 3])
a.*?
```

Here's the output showing the list:

```
a.T
a.__abs__
a.__add__
a.__and__
a.__array__
a.__array_finalize__
a.__array_function__
a.__array_interface__
a.__array_prepare__
a.__array_priority__
a.__array_struct__
a.__array_ufunc__
a.__array_wrap__
```

Figure 1.30: The output after running a.*?

5. Click an empty code cell in the **Tab Completion** section. Type **import** (including the space after) and then press the *Tab* key:

```
import <tab>
```

Tab completion can be used to do the following:

6. List the available modules when importing external libraries:

```
from numpy import <tab>
```

7. List the available modules of imported external libraries:

```
np.<tab>
```

8. Perform function and variable completion:

```
np.ar<tab>
sor<tab>([2, 3, 1])
myvar_1 = 5
myvar_2 = 6
my<tab>
```

Test each of these examples for yourself in the following cells:

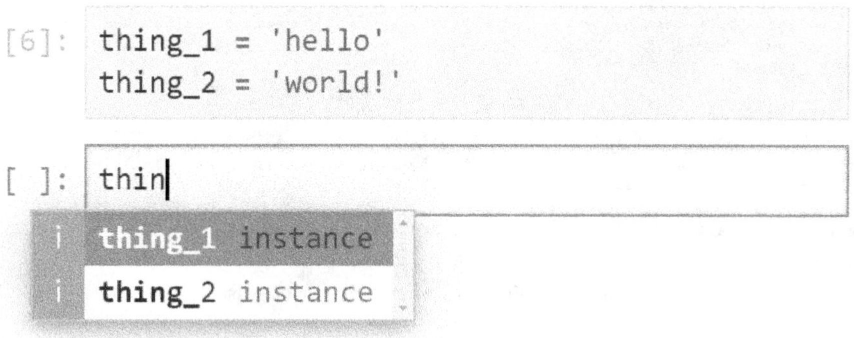

```
[6]:  thing_1 = 'hello'
      thing_2 = 'world!'
```

```
[ ]:  thin
      i  thing_1  instance
      i  thing_2  instance
```

Figure 1.31: An example of tab completion for variable names

NOTE

Tab completion is different in the JupyterLab and Jupyter Notebook platforms. The same commands may not work on both.

Tab completion can be especially useful when you need to know the available input arguments for a module, explore a new library, discover new modules, or simply speed up the workflow. They will save time writing out variable names or functions and reduce bugs from typos. Tab completion works so well that you may have difficulty coding Python in other editors after today.

9. List the available magic commands, as follows:

```
%lsmagic
```

The output is as follows:

```
Available line magics:
%alias %alias_magic %autoawait %autocall %automagic %autosave %bookmark %cat %cd %clear %c
olors %conda %config %connect_info %cp %debug %dhist %dirs %doctest_mode %ed %edit %env
%gui %hist %history %killbgscripts %ldir %less %lf %lk %ll %load %load_ext %loadpy %logo
ff %logon %logstart %logstate %logstop %ls %lsmagic %lx %macro %magic %man %matplotlib %
mkdir %more %mv %notebook %page %pastebin %pdb %pdef %pdoc %pfile %pinfo %pinfo2 %pip %
popd %pprint %precision %prun %psearch %psource %pushd %pwd %pycat %pylab %qtconsole %qui
ckref %recall %rehashx %reload_ext %rep %rerun %reset %reset_selective %rm %rmdir %run %s
ave %sc %set_env %store %sx %system %tb %time %timeit %unalias %unload_ext %who %who_ls
%whos %xdel %xmode

Available cell magics:
%%! %%HTML %%SVG %%bash %%capture %%debug %%file %%html %%javascript %%js %%latex %%markd
own %%perl %%prun %%pypy %%python %%python2 %%python3 %%ruby %%script %%sh %%svg %%sx %%
system %%time %%timeit %%writefile

Automagic is ON, % prefix IS NOT needed for line magics.
```

Figure 1.32: Jupyter magic functions

NOTE

If you're running JupyterLab, you will not see the preceding output. A list of magic functions, along with information about each, can be found here: https://ipython.readthedocs.io/en/stable/interactive/magics.html.

The percent signs, `%` and `%%`, are one of the basic features of Jupyter Notebook and are called magic commands. Magic commands starting with `%%` will apply to the entire cell, while magic commands starting with `%` will only apply to that line.

10. One example of a magic command that you will see regularly is as follows. This is used to display plots inline, which avoids you having to type `plt.show()` each time you plot something. You only need to execute it once at the beginning of the session:

```
%matplotlib inline
```

The timing functions are also very handy magic functions and come in two varieties: a standard timer (**%time** or **%%time**) and a timer that measures the average runtime of many iterations (**%timeit** and **%%timeit**). We'll see them being used here.

11. Declare the **a** variable, as follows:

```
a = [1, 2, 3, 4, 5] * int(1e5)
```

12. Get the runtime for the entire cell, as follows:

```
%%time
for i in range(len(a)):
    a[i] += 5
```

The output is as follows:

```
CPU times: user 68.8 ms, sys: 2.04 ms, total: 70.8 ms
Wall time: 69.6 ms
```

13. Get the runtime for one line:

```
%time a = [_a + 5 for _a in a]
```

The output is as follows:

```
CPU times: user 21.1 ms, sys: 2.6 ms, total: 23.7 ms
Wall time: 23.1 ms
```

14. Check the average results of multiple runs, as follows:

```
%timeit set(a)
```

The output is as follows:

```
4.72 ms ± 55.5 µs per loop (mean ± std. dev. of 7 runs, 100 loops
each)
```

Note the difference in use between one and two percent signs. Even when using a Python kernel (as you are currently doing), other languages can be invoked using magic commands. The built-in options include JavaScript, R, Perl, Ruby, and Bash. Bash is particularly useful as you can use Unix commands to find out where you are currently (**pwd**), see what's in the directory (**ls**), make new folders (**mkdir**), and write file contents (**cat/head/tail**).

> **NOTE**
>
> Notice how list comprehensions are quicker than loops in Python. This can be seen by comparing the wall time for the first and second cell, where the same calculation is done significantly faster with list comprehension. Please note that the step 15-18 are Linux-based commands. If you are working on other operating systems like Windows and MacOS, these steps might not work.

15. Write some text into a file in the working directory, print the directory's contents, print an empty line, and then write back the contents of the newly created file before removing it, as follows:

```
%%bash

echo "using bash from inside Jupyter!" > test-file.txt
ls
echo ""
cat test-file.txt
rm test-file.txt
```

The output is as follows:

```
%%bash

echo "using bash from inside Jupyter!" > test-file.txt
ls
echo ""
cat test-file.txt
rm test-file.txt
```

```
chapter_1_workbook.ipynb
test-file.txt

using bash from inside Jupyter!
```

Figure 1.33: Running a bash command in Jupyter

16. List the contents of the working directory with **ls**, as follows:

```
%ls
```

The output is as follows:

```
chapter_1_workbook.ipynb
```

17. List the path of the current working directory with **pwd**. Notice how we needed to use the **%%bash** magic function for **pwd**, but not for **ls**:

```
%%bash
pwd
```

The output is as follows:

```
/Users/alex/Documents/The-Applied-Data-Science-Workshop/chapter-01
```

18. There are plenty of external magic commands that can be installed. A popular one is **ipython-sql**, which allows for SQL code to be executed in cells.

    Jupyter magic functions can be installed the same way as PyPI Python libraries, using **pip** or **conda**. Open a new Terminal window and execute the following code to install **ipython-sql**:

    ```
    pip install ipython-sql
    ```

19. Run the **%load_ext sql** cell to load the external command into the Notebook.

    This allows connections to be made to remote databases so that queries can be executed (and thereby documented) right inside the Notebook.

20. Now, run the sample SQL query, as follows:

    ```
    %%sql sqlite://


    SELECT *
    FROM (
        SELECT 'Hello' as msg1, 'World!' as msg2
    );
    ```

    The output is as follows:

    Done.

msg1	msg2
Hello	World!

    Figure 1.34: Running a SQL query in Jupyter

    Here, we connected to the local **sqlite** source with **sqlite://**; however, this line could instead point to a specific database on a local or remote server. For example, a **.sqlite** database file on your desktop could be connected to with the line **%sql sqlite:////Users/alex/Desktop/db.sqlite**, where the username in this case is **alex** and the file is **db.sqlite**.

    After connecting, we execute a simple **SELECT** command to show how the cell has been converted to run SQL code instead of Python.

21. Earlier in this chapter, we went over the instructions for installing the **watermark** external library, which helps to document versioning in the Notebook. If you haven't installed it yet, then open a new window and run the following code:

```
pip install watermark
```

Once installed, it can be imported into any Notebook using **%load_ext watermark**. Then, it can be used to document library versions and system hardware.

22. Load and call the **watermark** magic function and call its docstring with the following command:

```
%load_ext watermark
%watermark?
```

The output is as follows:

```
Docstring:
::

  %watermark [-a AUTHOR] [-d] [-n] [-t] [-i] [-z] [-u] [-c CUSTOM_TIME]
             [-v] [-p PACKAGES] [-h] [-m] [-g] [-r] [-b] [-w] [-iv]

IPython magic function to print date/time stamps
and various system information.

optional arguments:
  -a AUTHOR, --author AUTHOR
                        prints author name
  -d, --date            prints current date as YYYY-mm-dd
  -n, --datename        prints date with abbrv. day and month names
  -t, --time            prints current time as HH-MM-SS
  -i, --iso8601         prints the combined date and time including the time
                        zone in the ISO 8601 standard with UTC offset
```

Figure 1.35: The docstring for watermark

Notice the various arguments that can be passed in when calling it, such as **-a** for author, **-v** for the Python version, **-m** for machine information, and **-d** for date.

23. Use the **watermark** library to add version information to the notebook, as follows:

> **NOTE**
>
> The code snippet shown here uses a backslash (\ ) to split the logic across multiple lines. When the code is executed, Python will ignore the backslash, and treat the code on the next line as a direct continuation of the current line.

```
%watermark -d -v -m -p \
requests,numpy,pandas,matplotlib,seaborn,sklearn
```

The output is as follows:

```
2020-02-09

CPython 3.7.5
IPython 7.10.1

requests 2.22.0
numpy 1.17.4
pandas 0.25.3
matplotlib 3.1.1
seaborn 0.9.0
sklearn 0.21.3

compiler   : Clang 4.0.1 (tags/RELEASE_401/final)
system     : Darwin
release    : 18.7.0
machine    : x86_64
processor  : i386
CPU cores  : 8
interpreter: 64bit
```

Figure 1.36: watermark output in the Notebook

> **NOTE**
>
> To access the source code for this specific section, please refer to https://packt.live/30KoAfu.
>
> You can also run this example online at https://packt.live/2Y49zTQ.

In this exercise, we looked at the core features of Jupyter, including tab completion and magic functions. You'll review these features and have a chance to test them out yourself in the activity at the end of this chapter.

CONVERTING A JUPYTER NOTEBOOK INTO A PYTHON SCRIPT

In this section, we'll learn how to convert a Jupyter Notebook into a Python script. This is equivalent to copying and pasting the contents of each code cell into a single **.py** file. The Markdown sections are also included as comments.

It can be beneficial to convert a Notebook into a **.py** file because the code is then available in plain text format. This can be helpful for version control— to see the difference in code between two versions of a Notebook, for example. It can also be a helpful trick for extracting chunks of code.

This conversion can be done from the Jupyter Dashboard (**File** -> **Download as**) or by opening a new Terminal window, navigating to the **chapter-02** folder, and executing the following:

```
jupyter nbconvert --to=python chapter_2_workbook.ipynb
```

The output is as follows:

```
(base) ➔ The-Applied-Data-Science-Workshop git:(master) × ls
LICENSE          chapter-01       chapter-03       chapter-05       chapter-4        figures          src
README.md        chapter-02       chapter-04       chapter-06       data             requirements.txt
(base) ➔ The-Applied-Data-Science-Workshop git:(master) × cd chapter-02
(base) ➔ chapter-02 git:(master) × ls -l
total 7520
-rw-r--r--  1 alex  staff  3848227 Mar 16 09:43 chapter_2_workbook.ipynb
(base) ➔ chapter-02 git:(master) × jupyter nbconvert --to=python chapter_2_workbook.ipynb
[NbConvertApp] Converting notebook chapter_2_workbook.ipynb to python
[NbConvertApp] Writing 9778 bytes to chapter_2_workbook.py
(base) ➔ chapter-02 git:(master) × ls -l
total 7544
-rw-r--r--  1 alex  staff  3848227 Mar 16 09:43 chapter_2_workbook.ipynb
-rw-r--r--  1 alex  staff     9780 Mar 16 09:48 chapter_2_workbook.py
(base) ➔ chapter-02 git:(master) × head chapter_2_workbook.py
#!/usr/bin/env python
# coding: utf-8

# <img src="../src/packt-banner.png" alt="">

# # Chapter 2: Data Exploration with Jupyter
#
#
# We get our hands on some data and work through an exploration analysis, where we compute some interesting
  and informative metrics and visualizations.
#
(base) ➔ chapter-02 git:(master) × ▮
```

Figure 1.37: Converting a Notebook into a script (.py) file

Note that we are using the next chapter's Notebook for this example.

Another benefit of converting Notebooks into **.py** files is that, when you want to determine the Python library requirements for a Notebook, the **pipreqs** tool will do this for us, and export them into a **requirements.txt** file. This tool can be installed by running the following command:

```
pip install pipreqs
```

You might require root privileges for this.

This command is called from outside the folder containing your **.py** files. For example, if the **.py** files are inside a folder called **chapter-02**, you could do the following:

```
pipreqs chapter-02/
```

The output is as follows:

```
(base) ➜ The-Applied-Data-Science-Workshop git:(master) × pipreqs chapter-02
INFO: Successfully saved requirements file in chapter-02/requirements.txt
(base) ➜ The-Applied-Data-Science-Workshop git:(master) × cat chapter-02/requirements.txt
pandas==0.25.3
numpy==1.17.4
matplotlib==3.1.1
requests==2.22.0
seaborn==0.9.0
beautifulsoup4==4.8.2
scikit_learn==0.22.2.post1
(base) ➜ The-Applied-Data-Science-Workshop git:(master) ×
```

Figure 1.38: Using pipreqs to generate a requirements.txt file

The resulting **requirements.txt** file for **chapter_2_workbook.ipynb** will look similar to the following:

```
cat chapter-02/requirements.txt
matplotlib==3.1.1
seaborn==0.9.0
numpy==1.17.4
pandas==0.25.3
requests==2.22.0
beautifulsoup4==4.8.1
scikit_learn==0.22
```

PYTHON LIBRARIES

Having now seen all the basics of Jupyter Notebooks, and even some more advanced features, we'll shift our attention to the Python libraries we'll be using in this book.

Libraries, in general, extend the default set of Python functions. Some examples of commonly used standard libraries are **datetime**, **time**, **os**, and **sys**. These are called standard libraries because they are included with every installation of Python.

For data science with Python, the most heavily relied upon libraries are external, which means they do not come as standard with Python.

The external data science libraries we'll be using in this book are **numpy**, **pandas**, **seaborn**, **matplotlib**, **scikit-learn**, **requests**, and **bokeh**.

> **NOTE**
>
> It's a good idea to import libraries using industry standards—for example, **import numpy as np**. This way, your code is more readable. Try to avoid doing things such as **from numpy import ***, as you may unwittingly overwrite functions. Furthermore, it's often nice to have modules linked to the library via a dot (.) for code readability.

Let's briefly introduce each:

- **numpy** offers multi-dimensional data structures (arrays) that operations can be performed on. This is far quicker than standard Python data structures (such as lists). This is done in part by performing operations in the background using C. NumPy also offers various mathematical and data manipulation functions.

- **pandas** is Python's answer to the R DataFrame. It stores data in 2D tabular structures where columns represent different variables and rows correspond to samples. pandas provides many handy tools for data wrangling, such as filling in **NaN** entries and computing statistical descriptions of the data. Working with pandas DataFrames will be a big focus of this book.

- **matplotlib** is a plotting tool inspired by the MATLAB platform. Those familiar with R can think of it as Python's version of **ggplot**. It's the most popular Python library for plotting figures and allows for a high level of customization.

- **seaborn** works as an extension of **matplotlib**, where various plotting tools that are useful for data science are included. Generally speaking, this allows for analysis to be done much faster than if you were to create the same things manually with libraries such as **matplotlib** and **scikit-learn**.

- **scikit-learn** is the most commonly used machine learning library. It offers top-of-the-line algorithms and a very elegant API where models are instantiated and then fit with data. It also provides data processing modules and other tools that are useful for predictive analytics.

- **requests** is the go-to library for making HTTP requests. It makes it straightforward to get HTML from web pages and interface with APIs. For parsing HTML, many choose **BeautifulSoup4**, which we'll cover in *Chapter 6, Web Scraping with Jupyter Notebooks*.

We'll start using these libraries in the next chapter.

ACTIVITY 1.01: USING JUPYTER TO LEARN ABOUT PANDAS DATAFRAMES

We are going to be using pandas heavily in this book. In particular, any data that's loaded into our Notebooks will be done using pandas. The data will be contained in a DataFrame object, which can then be transformed and saved back to disk afterward. These DataFrames offer convenient methods for running calculations over the data for exploration, visualization, and modeling.

In this activity, you'll have the opportunity to use pandas DataFrames, along with the Jupyter features that have been discussed in this section. Follow these steps to complete this activity:

1. Start up one of the platforms for running Jupyter Notebooks and open it in your web browser by copying and pasting the URL, as prompted in the Terminal.

   > **NOTE**
   >
   > While completing this activity, you will need to use many cells in the Notebook. Please insert new cells as required.

2. Import the pandas and NumPy libraries and assign them to the **pd** and **np** variables, respectively.

3. Pull up the docstring for **pd.DataFrame**. Scan through the Parameters section and read the Examples section.

4. Create a dictionary with **fruit** and **score** keys, which correspond to list values with at least three items in each. Ensure that you give your dictionary a suitable name (note that in Python, a dictionary is a collection of values); for example, `{"fruit": ["apple", ...], "score": [8, ...]}`.

5. Use this dictionary to create a DataFrame. You can do this using `pd.DataFrame(data=name of dictionary)`. Assign it to the **df** variable.

6. Display this DataFrame in the Notebook.

7. Use tab completion to pull up a list of functions available for **df**.

8. Pull up the docstring for the **sort_values** DataFrame function and read through the Examples section.

9. Sort the DataFrame by score in descending order. Try to see how many times you can use tab completion as you write the code.

10. Use the **timeit** magic function to test how long this sorting operation takes.

> **NOTE**
>
> The detailed steps for this activity, along with the solutions, are presented on page 262.

SUMMARY

In this chapter, we've gone over the basics of using Jupyter Notebooks for data science. We started by exploring the platform and finding our way around the interface. Then, we discussed the most useful features, which include tab completion and magic functions. Finally, we introduced the Python libraries we'll be using in this book.

As we'll see in the coming chapters, these libraries offer high-level abstractions that allow data science to be highly accessible with Python. This includes methods for creating statistical visualizations, building data cleaning pipelines, and training models on millions of data points and beyond.

While this chapter focused on the basics of Jupyter platforms, the next chapter is where the real data science begins. The remainder of this book is very interactive, and in *Chapter 3, Preparing Data for Predictive Modeling*, we'll perform an analysis of housing data using Jupyter Notebook and the Seaborn plotting library.

2

DATA EXPLORATION WITH
JUPYTER

OVERVIEW

In this chapter, we'll finally get our hands on some data and work through an exploratory analysis, where we'll compute some informative metrics and visualizations. By the end of this chapter, you will be able to use the pandas Python library to load tabular data and run calculations on it, and the **seaborn** Python library to create visualizations.

INTRODUCTION

So far, we have taken a glance at the data science ecosystem and jumped into learning about Jupyter, the tool that we'll be using throughout this book for our coding exercises and activities. Now, we'll shift our focus away from learning about Jupyter and start actually using it for analysis.

Data visualization and **exploration** are important steps in the data science process. This is how you can learn about your data and make sure you understand it completely. **Visualizations** can be used as a means of discovering unusual records in datasets and presenting that information to others.

In addition to understanding and gaining fundamental trust in data, your analysis may lead to the discovery of patterns and insights in the data. In some cases, these patterns can prompt further research and ultimately be very beneficial to your business.

Applied knowledge of a high-level programming language such as Python or R will make datasets accessible to you, from top-level aggregations to granular details. However, it's also possible to learn a lot from data with tools that are easier to pick up and use, such as Tableau or Microsoft Power BI.

In addition to learning about the tools to create them, it's important to have a conceptual understanding of different types of visualizations and their uses. Similarly, there are a handful of important techniques relating to data exploration, such as aggregation and filtering for outliers or missing samples.

In this chapter, we'll start out by learning about some of the basics of working with datasets in Jupyter by using pandas DataFrames. Then, we'll learn about exploring datasets with the Seaborn visualization library and do basic modeling with scikit learn.

OUR FIRST ANALYSIS – THE BOSTON HOUSING DATASET

The dataset we'll be looking at in this section is the so-called Boston Housing dataset. It contains US census data concerning houses in various areas around the city of Boston. Each sample corresponds to a unique area and has about a dozen measures. We should think of samples as rows and measures as columns. This data was first published in 1978 and is quite small, containing only about 500 samples.

Now that we know something about the context of the dataset, let's decide on a rough plan for the exploration and analysis stages. If applicable, this plan will accommodate the relevant questions under study. In this case, the goal is not to answer a question, but to show Jupyter in action and illustrate some basic data analysis methods.

Our general approach to this analysis will be to do the following:

- Load the data into Jupyter using a pandas DataFrame

- Quantitatively understand the features

- Look for patterns and generate questions

- Answer the questions to the problems

Before we start this analysis, let's take a moment to set up our notebook environment.

EXERCISE 2.01: IMPORTING DATA SCIENCE LIBRARIES AND SETTING UP THE NOTEBOOK PLOTTING ENVIRONMENT

In this exercise, we'll set the stage so that we're ready to load data into the notebook. Perform the following steps to complete this exercise:

1. If you haven't done so already, start up one of the following platforms for running Jupyter Notebooks:

    JupyterLab (run **jupyter lab**)

    Jupyter Notebook (run **jupyter notebook**)

2. Then, open your chosen platform in a web browser by copying and pasting the URL, as prompted in the Terminal.

3. Import **pandas** and pull up the **docstring** for a pandas DataFrame, the main data structure you will use to store tables and run calculations on them. This should be familiar to you from *Activity 1.01, Using Jupyter to Learn about Pandas DataFrames*, from *Chapter 1, Introduction to Jupyter Notebooks*, where you learned about some basics of DataFrames:

```
import pandas as pd
pd.DataFrame?
```

The output is as follows:

```
Init signature: pd.DataFrame(data=None, index=None, columns=None, dtype=None, copy=False)
Docstring:
Two-dimensional size-mutable, potentially heterogeneous tabular data
structure with labeled axes (rows and columns). Arithmetic operations
align on both row and column labels. Can be thought of as a dict-like
container for Series objects. The primary pandas data structure.

Parameters
----------
data : ndarray (structured or homogeneous), Iterable, dict, or DataFrame
    Dict can contain Series, arrays, constants, or list-like objects

    .. versionchanged :: 0.23.0
       If data is a dict, column order follows insertion-order for
       Python 3.6 and later.

    .. versionchanged :: 0.25.0
       If data is a list of dicts, column order follows insertion-order
       for Python 3.6 and later.
```

Figure 2.1: The docstring for pd.DataFrame

4. Import **seaborn** and pull up examples of the plot functions you have access to:

```
import seaborn as sns
sns.*?
```

You'll see familiar charts in the list, such as barplot, boxplot, and scatterplot, as follows:

```
sns.FacetGrid
sns.JointGrid
sns.PairGrid
sns.__builtins__
sns.__cached__
sns.__class__
sns.__delattr__
sns.__dict__
sns.__dir__
sns.__doc__
```

Figure 2.2: The output of running sns.*?

5. Load the library dependencies and set up the plotting environment for the notebook.

> **NOTE**
>
> You can type these libraries into new cells and play with the Jupyter tab completion features.

The code for importing the libraries is as follows:

```
# Common standard libraries
import datetime
import time
import os
# Common external libraries

import pandas as pd
import numpy as np
import sklearn # scikit-learn
import requests
from bs4 import BeautifulSoup
# Visualization libraries

%matplotlib inline
import matplotlib.pyplot as plt
import seaborn as sns
```

Libraries do not need to be imported at the top of the notebook, but it's good practice to do this in order to summarize the dependencies in one place. Sometimes, though, it makes sense to load things midway through the notebook, and that is completely fine.

> **NOTE**
>
> To access the source code for this specific section, please refer to https://packt.live/2N1riF2.
>
> You can also run this example online at https://packt.live/37DzuVK.

6. Set the plot's appearance, as follows:

```
%config InlineBackend.figure_format='retina'
sns.set() # Revert to matplotlib defaults
plt.rcParams['figure.figsize'] = (9, 6)
plt.rcParams['axes.labelpad'] = 10
sns.set_style("darkgrid")
```

> **NOTE**
>
> It's easy to run into issues of non-reproducibility when working in a notebook by running cells out of order. For this reason, you can keep an eye on the cell run count, which is listed to the left of the cell in square brackets and increments by one each time a cell is run.
>
> Consider, for example, if you import pandas (**import pandas as pd**) halfway down the notebook. Since Jupyter notebooks have shared memory between cells, you can use pandas in any cell, even those above it. Now, say you did this and used pandas to write some code above the import cell. In this case, if you were to rerun the notebook from scratch (or send to someone else to run), then it will throw a **NameError** exception and stop executing when it hits that cell. This will happen because the notebook is attempting to execute pandas code before importing the library.
>
> A similar issue can appear when transforming data, such as rerunning charts on data that was transformed later in the notebook. In this case, the solution is to be careful about the order in which you run things. Try to rerun your notebook from scratch occasionally if time permits so that you can catch errors like this before they creep too far into your analysis.

Now that we've loaded our external libraries, we can move on to the analysis. Please keep the notebook open and carry on to the next exercise to continue running through the notebook.

LOADING THE DATA INTO JUPYTER USING A PANDAS DATAFRAME

Often, data is stored in tables, which means it can be saved as a **comma-separated variable (CSV)** file. This format, and many others, can be read into Python as a DataFrame object using the pandas library. Other common formats include **tab-separated variable (TSV)**, **Structured Query Language (SQL)** tables, and **JavaScript Object Notation (JSON)** data structures. Indeed, pandas has support for all of these.

We won't need to worry about loading data like this right now, since our first dataset is available directly through scikit-learn; however, we will see some examples of loading and saving these types of data structures later in this book.

> **NOTE**
>
> An important step after loading data for analysis is to ensure that it's clean. For example, we would generally need to deal with missing data and ensure that all the columns have the correct data types. The dataset we'll be using in this section has already been cleaned, so we won't need to worry about this. However, we'll see an example of data that requires cleaning in *Chapter 3, Preparing Data for Predictive Modeling*, and explore techniques for dealing with it.

EXERCISE 2.02: LOADING THE BOSTON HOUSING DATASET

To get started with the Boston Housing dataset, you will load the table into a pandas DataFrame and look at its fields and some summary statistics. Perform the following steps to complete this exercise:

1. Load the Boston Housing dataset from the `sklearn.datasets` module using the `load_boston` method:

```
from sklearn import datasets
boston = datasets.load_boston()
```

2. Check the data structure type, as follows:

```
type(boston)
```

The output is as follows:

```
sklearn.utils.Bunch
```

The output indicates that it's a scikit-learn **Bunch** object, but you still need to get some more information about this to understand what you are dealing with.

3. Import the base object from scikit-learn **utils** and print the docstring in your notebook:

```
from sklearn.utils import Bunch
Bunch?
```

The output is as follows:

```
Init signature: Bunch(**kwargs)
Docstring:
Container object for datasets

Dictionary-like object that exposes its keys as attributes.

>>> b = Bunch(a=1, b=2)
>>> b['b']
2
>>> b.b
2
>>> b.a = 3
>>> b['a']
3
>>> b.c = 6
>>> b['c']
6
File:           /anaconda3/lib/python3.7/site-packages/sklearn/utils/__init__.py
Type:           type
Subclasses:
```

Figure 2.3: The docstring for sklearn.utils.Bunch

4. Print the field names (that is, the keys to the dictionary) as follows:

```
boston.keys()
```

The output is as follows:

```
dict_keys(['data', 'target', 'feature_names', 'DESCR', 'filename'])
```

5. Print the dataset description contained in **boston['DESCR']** as follows:

```
print(boston['DESCR'])
```

Note that, in this call, you want to explicitly print the field value so that the notebook renders the content in a more readable format than the string representation (that is, if we just type **boston['DESCR']** without wrapping it in a **print** statement). By doing this, we can see the dataset information as we summarized it previously:

```
.. _boston_dataset:

Boston house prices dataset
---------------------------

**Data Set Characteristics:**

    :Number of Instances: 506

    :Number of Attributes: 13 numeric/categorical predictive. Median Value (attribute 14) is usually
the target.

    :Attribute Information (in order):
        - CRIM     per capita crime rate by town
        - ZN       proportion of residential land zoned for lots over 25,000 sq.ft.
        - INDUS    proportion of non-retail business acres per town
        - CHAS     Charles River dummy variable (= 1 if tract bounds river; 0 otherwise)
        - NOX      nitric oxides concentration (parts per 10 million)
        - RM       average number of rooms per dwelling
        - AGE      proportion of owner-occupied units built prior to 1940
        - DIS      weighted distances to five Boston employment centres
        - RAD      index of accessibility to radial highways
        - TAX      full-value property-tax rate per $10,000
        - PTRATIO  pupil-teacher ratio by town
        - B        1000(Bk - 0.63)^2 where Bk is the proportion of blacks by town
        - LSTAT    % lower status of the population
        - MEDV     Median value of owner-occupied homes in $1000's
```

Figure 2.4: Description of the Boston dataset

Briefly read through the feature descriptions and/or describe them yourself. For the purposes of this book, the most important fields to understand are **RM**, **AGE**, **LSTAT**, and **MEDV**. Of particular importance, here, are the feature descriptions (under **Attribute Information**). You will use this as a reference during your analysis.

6. Now, create a pandas DataFrame that contains the data. This is beneficial for a few reasons: all of our data will be contained in one object, there are useful and computationally efficient DataFrame methods we can use, and other libraries, such as seaborn, have tools that integrate nicely with DataFrames.

   In this case, you will create your DataFrame with the standard constructor method. As a reminder of how this is done, pull up the docstring one more time:

```
import pandas as pd
pd.DataFrame?
```

The output is as follows:

```
Init signature: pd.DataFrame(data=None, index=None, columns=None, dtype=None, copy=False)
Docstring:
Two-dimensional size-mutable, potentially heterogeneous tabular data
structure with labeled axes (rows and columns). Arithmetic operations
align on both row and column labels. Can be thought of as a dict-like
container for Series objects. The primary pandas data structure.

Parameters
----------
data : ndarray (structured or homogeneous), Iterable, dict, or DataFrame
    Dict can contain Series, arrays, constants, or list-like objects

    .. versionchanged :: 0.23.0
       If data is a dict, column order follows insertion-order for
       Python 3.6 and later.

    .. versionchanged :: 0.25.0
       If data is a list of dicts, column order follows insertion-order
       for Python 3.6 and later.
```

Figure 2.5: The docstring for pd.DataFrame

The docstring reveals the DataFrame input parameters. You want to feed in **boston['data']** for the data and use **boston['feature_names']** for the headers.

7. Print the data, as follows:

```
boston['data']
```

The output is as follows:

```
array([[6.3200e-03, 1.8000e+01, 2.3100e+00, ..., 1.5300e+01, 3.9690e+02,
        4.9800e+00],
       [2.7310e-02, 0.0000e+00, 7.0700e+00, ..., 1.7800e+01, 3.9690e+02,
        9.1400e+00],
       [2.7290e-02, 0.0000e+00, 7.0700e+00, ..., 1.7800e+01, 3.9283e+02,
        4.0300e+00],
       ...,
       [6.0760e-02, 0.0000e+00, 1.1930e+01, ..., 2.1000e+01, 3.9690e+02,
        5.6400e+00],
       [1.0959e-01, 0.0000e+00, 1.1930e+01, ..., 2.1000e+01, 3.9345e+02,
        6.4800e+00],
       [4.7410e-02, 0.0000e+00, 1.1930e+01, ..., 2.1000e+01, 3.9690e+02,
        7.8800e+00]])
```

Figure 2.6: Printing the data

8. Print the shape of the data, as follows:

```
boston['data'].shape
```

The output is as follows:

```
(506, 13)
```

9. Print the feature names, as follows:

```
boston['feature_names']
```

The output is as follows:

```
array(['CRIM', 'ZN', 'INDUS', 'CHAS', 'NOX', 'RM', 'AGE', 'DIS',
       'RAD', 'TAX', 'PTRATIO', 'B', 'LSTAT'], dtype='<U7')
```

Looking at the output, you can see that your data is in a 2D NumPy array. Running the **boston['data'].shape** command returns the length (number of samples) and the number of features as the first and second outputs, respectively.

10. Load the data into a pandas DataFrame, **df**, by running the following code:

```
# Load the data
df = pd.DataFrame(data=boston['data'], \
                  columns=boston['feature_names'],)
```

In machine learning, the variable that is being modeled is called the **target variable**; it's what you are trying to predict, given the features. For this dataset, the suggested target is **MEDV**, which is the median house value in thousands of dollars.

11. Check the shape of the target, as follows:

```
boston['target'].shape
```

The output is **(506,)**.

You can see that it has the same length as the features, which is expected. Therefore, it can be added as a new column to the DataFrame.

12. Add the target variable to **df**, as follows:

```
df['MEDV'] = boston['target']
```

13. Move the target variable to the front of **df**, as follows:

```
y = df['MEDV'].copy()
del df['MEDV']
df = pd.concat((y, df), axis=1)
```

This is done to distinguish the target from our features by storing it at the front of our DataFrame.

Here, you've introduced a dummy variable, **y**, to hold a copy of the target column before we remove it from the DataFrame. Then, you used the pandas concatenation function to combine it with the remaining DataFrame along the 1st axis (as opposed to the 0th axis, which combines rows).

> ### NOTE
>
> You will often see dot notation used to reference DataFrame columns. For example, previously, we could have done **y = df.MEDV.copy()**. While this can work well for accessing column data, it cannot be used generally in place of the square bracket notation, such as when you have column names with spaces. Additionally, you cannot delete columns using dot notation. For example, **del df.MEDV** would raise an error. The correct way to delete the **MEDV** column would be to run **del df['MEDV']**.

14. Implement **df.head()** or **df.tail()** to glimpse the data and **len(df)** to verify that the number of samples is what we expect. Run the next few cells to see the head, tail, and length of **df**:

```
[21]: df.head()
```

	MEDV	CRIM	ZN	INDUS	CHAS	NOX	RM	AGE	DIS	RAD	TAX	PTRATIO	B	LSTAT
0	24.0	0.00632	18.0	2.31	0.0	0.538	6.575	65.2	4.0900	1.0	296.0	15.3	396.90	4.98
1	21.6	0.02731	0.0	7.07	0.0	0.469	6.421	78.9	4.9671	2.0	242.0	17.8	396.90	9.14
2	34.7	0.02729	0.0	7.07	0.0	0.469	7.185	61.1	4.9671	2.0	242.0	17.8	392.83	4.03
3	33.4	0.03237	0.0	2.18	0.0	0.458	6.998	45.8	6.0622	3.0	222.0	18.7	394.63	2.94
4	36.2	0.06905	0.0	2.18	0.0	0.458	7.147	54.2	6.0622	3.0	222.0	18.7	396.90	5.33

```
[22]: df.tail(10)
```

	MEDV	CRIM	ZN	INDUS	CHAS	NOX	RM	AGE	DIS	RAD	TAX	PTRATIO	B	LSTAT
496	19.7	0.28960	0.0	9.69	0.0	0.585	5.390	72.9	2.7986	6.0	391.0	19.2	396.90	21.14
497	18.3	0.26838	0.0	9.69	0.0	0.585	5.794	70.6	2.8927	6.0	391.0	19.2	396.90	14.10
498	21.2	0.23912	0.0	9.69	0.0	0.585	6.019	65.3	2.4091	6.0	391.0	19.2	396.90	12.92
499	17.5	0.17783	0.0	9.69	0.0	0.585	5.569	73.5	2.3999	6.0	391.0	19.2	395.77	15.10
500	16.8	0.22438	0.0	9.69	0.0	0.585	6.027	79.7	2.4982	6.0	391.0	19.2	396.90	14.33
501	22.4	0.06263	0.0	11.93	0.0	0.573	6.593	69.1	2.4786	1.0	273.0	21.0	391.99	9.67
502	20.6	0.04527	0.0	11.93	0.0	0.573	6.120	76.7	2.2875	1.0	273.0	21.0	396.90	9.08
503	23.9	0.06076	0.0	11.93	0.0	0.573	6.976	91.0	2.1675	1.0	273.0	21.0	396.90	5.64
504	22.0	0.10959	0.0	11.93	0.0	0.573	6.794	89.3	2.3889	1.0	273.0	21.0	393.45	6.48
505	11.9	0.04741	0.0	11.93	0.0	0.573	6.030	80.8	2.5050	1.0	273.0	21.0	396.90	7.88

```
[23]: len(df)
```

```
[23]: 506
```

Figure 2.7: The first 10 and last 10 rows of the Boston Housing dataset

15. Verify that the number of samples is what you expect, as follows:

```
len(df)
```

The output is **506**.

Each row is labeled with an index value, as seen in bold on the left-hand side of the table. By default, these are a set of integers starting at 0 and incrementing by one for each row.

16. Check the data type contained within each column, as follows:

```
df.dtypes
```

The output is as follows:

```
MEDV        float64
CRIM        float64
ZN          float64
INDUS       float64
CHAS        float64
NOX         float64
RM          float64
AGE         float64
DIS         float64
RAD         float64
TAX         float64
PTRATIO     float64
B           float64
LSTAT       float64
dtype:  object
```

Figure 2.8: Data types of each column in the Boston Housing dataset

For this dataset, you can see that every field is a float, including the target. Looking at a sample of the target (**MEDV**) specifically, you can clearly see that it's distributed over a continuous range, as would be expected of a float data type. This makes sense, considering the field represents the median house value of each record. Given this information, you can determine that predicting this target variable is a regression problem.

17. You can run **df.isnull()** to clean the dataset as pandas automatically sets missing data as NaN values. To get the number of NaN values per column, execute the following code:

```
df.isnull().sum()
```

The output is as follows:

```
MEDV        0
CRIM        0
ZN          0
INDUS       0
CHAS        0
NOX         0
RM          0
AGE         0
DIS         0
RAD         0
TAX         0
PTRATIO     0
B           0
LSTAT       0
dtype: int64
```

Figure 2.9: Number of missing records in each column of the Boston Housing dataset

`df.isnull()` returns a Boolean frame of the same length as `df`.

For this dataset, you can see that there are no NaN values, which means you have no immediate work to do in terms of cleaning the data and can move on.

18. Remove some columns, as follows:

```
for col in ['ZN', 'NOX', 'RAD', 'PTRATIO', 'B']:
    del df[col]
```

This is done to simplify the analysis. We will focus on the remaining columns in more detail when we dive into exploring the data.

> **NOTE**
>
> To access the source code for this specific section, please refer to https://packt.live/2N1riF2.
>
> You can also run this example online at https://packt.live/37DzuVK.

DATA EXPLORATION

Since this is an entirely new dataset that we've never seen before, the first goal here is to understand the data. We've already seen the textual description of the data, which is important for qualitative understanding. Now, we'll compute a quantitative description.

EXERCISE 2.03: ANALYZING THE BOSTON HOUSING DATASET

In this exercise, you will learn more about the dataset from a top-down perspective, starting with summary metrics and then digging into more details of particular columns. You will explore relationships between the various fields, plot out visualizations to increase our understanding, and look into questions that arise. Perform the following steps to complete this exercise:

1. Check some statistics of the DataFrame, as follows:

```
df.describe().T
```

The output is as follows:

	count	mean	std	min	25%	50%	75%	max
MEDV	506.0	22.532806	9.197104	5.00000	17.025000	21.20000	25.000000	50.0000
CRIM	506.0	3.613524	8.601545	0.00632	0.082045	0.25651	3.677083	88.9762
INDUS	506.0	11.136779	6.860353	0.46000	5.190000	9.69000	18.100000	27.7400
CHAS	506.0	0.069170	0.253994	0.00000	0.000000	0.00000	0.000000	1.0000
RM	506.0	6.284634	0.702617	3.56100	5.885500	6.20850	6.623500	8.7800
AGE	506.0	68.574901	28.148861	2.90000	45.025000	77.50000	94.075000	100.0000
DIS	506.0	3.795043	2.105710	1.12960	2.100175	3.20745	5.188425	12.1265
TAX	506.0	408.237154	168.537116	187.00000	279.000000	330.00000	666.000000	711.0000
LSTAT	506.0	12.653063	7.141062	1.73000	6.950000	11.36000	16.955000	37.9700

Figure 2.10: Summary statistics for the Boston Housing dataset

The preceding command computes various properties, including the mean, standard deviation, minimum, and maximum for each column. This table gives us a high-level idea of how everything is distributed. Note that you have taken the transform of the result by adding a `.T` suffix to the output; this swaps the rows and columns.

Going forward with the analysis, you will specify a set of columns to focus on.

2. Define a specific set of columns, as follows:

```
cols = ['RM', 'AGE', 'TAX', 'LSTAT', 'MEDV']
```

3. Display this subset as the columns of the DataFrame, as follows:

```
df[cols].tail()
```

The output is as follows:

	RM	AGE	TAX	LSTAT	MEDV
501	6.593	69.1	273.0	9.67	22.4
502	6.120	76.7	273.0	9.08	20.6
503	6.976	91.0	273.0	5.64	23.9
504	6.794	89.3	273.0	6.48	22.0
505	6.030	80.8	273.0	7.88	11.9

Figure 2.11: A subset of columns from the Boston Housing dataset

Recall what each of these columns represents. From the dataset documentation, you have the following:

RM: Average number of rooms per dwelling

AGE: Proportion of owner-occupied units built prior to 1940

TAX: Full-value property tax rate per $10,000

LSTAT: Percentage of the population that is classified as "low status"

MEDV: Median value of owner-occupied homes in $1000s

To look for patterns in this data, you can start by calculating the pairwise correlations using **pd.DataFrame.corr**.

4. Calculate the pairwise correlations for your selected columns, as follows:

```
df[cols].corr()
```

The output of this command is as follows:

	RM	AGE	TAX	LSTAT	MEDV
RM	1.000000	-0.240265	-0.292048	-0.613808	0.695360
AGE	-0.240265	1.000000	0.506456	0.602339	-0.376955
TAX	-0.292048	0.506456	1.000000	0.543993	-0.468536
LSTAT	-0.613808	0.602339	0.543993	1.000000	-0.737663
MEDV	0.695360	-0.376955	-0.468536	-0.737663	1.000000

Figure 2.12: Pairwise correlation scores between select columns

This resulting table shows the correlation score between each pair of columns. Large positive scores indicate a strong positive (that is, in the same direction) correlation. As expected, you can see maximum values of 1 on the diagonal.

By default, pandas calculates the standard correlation coefficient for each pair of columns, which is also called the **Pearson coefficient**. This is defined as the covariance between two variables, divided by the product of their standard deviations:

$$\rho_{X,\,Y} = \frac{cov(X,\,Y)}{\sigma_X \sigma_Y}$$

Figure 2.13: The Pearson correlation coefficient equation

Here, you should think of X as one column, and Y as another. The standard deviations are calculated in the usual way, by summing up the squared differences between each data point and the average for that column.

The covariance, in turn, is defined as follows:

$$cov(x,\,y) = \frac{\sum_{i=1}^{n} (x_i - \bar{x})(y_i - \bar{y})}{n-1}$$

Figure 2.14: The covariance equation

Here, n is the number of records (that is, the number of rows in the table), x_i and y_i correspond to the individual values of each record being summed over, and $\bar{x}$ and $\bar{y}$ correspond to the average values of the records for columns X and Y, respectively.

Returning to the analysis, visualize the correlation coefficients that you calculated previously, using a heatmap. This will produce a better result for you to look at, rather than having to strain your eyes to interpret the preceding table.

5. Import the libraries that are required to plot a heatmap and set the appearance settings as follows:

```
# Visualization libraries
import matplotlib.pyplot as plt
%matplotlib inline
import seaborn as sns


# Setting plot appearance
%config InlineBackend.figure_format='retina'
sns.set() # Revert to matplotlib defaults
plt.rcParams['figure.figsize'] = (9, 6)
plt.rcParams['axes.labelpad'] = 10
sns.set_style("darkgrid")
```

6. To create the heatmap using Seaborn, execute the following code:

```
ax = sns.heatmap(df[cols].corr(), \
                 cmap=sns.cubehelix_palette(20, \
                                            light=0.95, \
                                            dark=0.15),)
ax.xaxis.tick_top() # move labels to the top

plt.savefig('../figures/chapter-2-boston-housing-corr.png', \
            bbox_inches='tight', dpi=300,)
```

> **NOTE**
>
> To save the image using the preceding code, you will need to ensure you have a **figures** folder set up in your working directory. Alternatively, you can edit the path in the code to save the image to a different folder of your choice.

The following screenshot shows the heatmap:

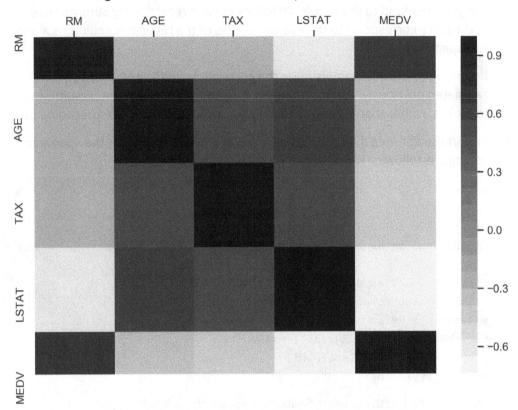

Figure 2.15: Heatmap of correlation scores

Call **sns.heatmap** and pass the pairwise correlation matrix as input. You will use a custom color palette here to override the Seaborn default. The function returns a **matplotlib.axes** object, which is referenced by the **ax** variable.

The final figure is then saved as a high-resolution PNG to the **figures** folder.

> **NOTE**
>
> Each chart is exported as a PNG file using the **plt.savefig** function. We set the **bbox_inches='tight'** and **dpi=300** argument in order to save a high-quality image.

For the final step in our dataset exploration exercise, we'll visualize our data using seaborn's **pairplot** function.

7. Visualize the DataFrame using Sseaborn's **pairplot** function, as follows:

```
sns.pairplot(df[cols], plot_kws={'alpha': 0.6}, \
             diag_kws={'bins': 30},)
plt.savefig('../figures/chapter-2-boston-housing-pairplot.png', \
            bbox_inches='tight', dpi=300,)
```

Here's the output of the visualization:

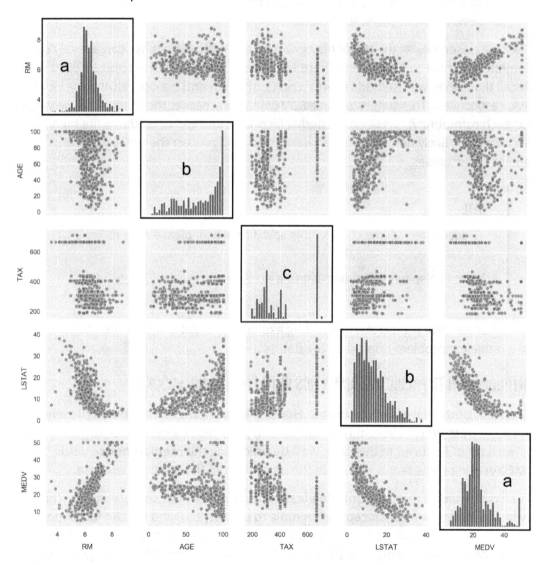

Figure 2.16: Pairplot visualization of the Boston Housing dataset

Looking at the histograms on the diagonal, you can see the following:

- **a**: **RM** and **MEDV** have the closest shape to normal distributions.

- **b**: **AGE** is skewed to the left and **LSTAT** is skewed to the right (this may seem counterintuitive, but skew is defined in terms of where the mean is positioned in relation to the max).

- **c**: For **TAX**, we find that a large amount of the distribution is around 700. This is also evident from the scatter plots.

Taking a closer look at the **MEDV** histogram on the bottom right, we can actually see something similar to **TAX**, where there is a large upper-limit bin around $50,000. Recall that, when we used `df.describe()`, the min and max of **MEDV** were 5k and 50k, respectively. This suggests that **MEDV** (which represents the median house values for each community) has been capped at 5k on the low end and 50k on the high end. This was likely done for ethical reasons, to help protect the privacy of the outlier communities.

> **NOTE**
>
> To access the source code for this specific section, please refer to https://packt.live/2N1riF2.
>
> You can also run this example online at https://packt.live/37DzuVK.

Now that we have a good start regarding exploring the data, let's turn our attention to a modeling problem offered by the dataset.

INTRODUCTION TO PREDICTIVE ANALYTICS WITH JUPYTER NOTEBOOKS

Continuing our analysis of the Boston Housing dataset, we can see that it presents us with a regression problem. In regression, we try to predict a numerical target variable, given a set of features. In this case, we'll be predicting the **median house value (MEDV)** using some features seen in the pairplot from the previous exercise.

We'll train models that take only one feature as input to make this prediction. This way, the models will be conceptually simple to understand, and we can focus more on the technical details of the scikit-learn API. Then, in the next chapter, you'll be more comfortable dealing with the relatively complicated models we are going to train there.

EXERCISE 2.04: TRAINING LINEAR MODELS WITH SEABORN AND SCIKIT-LEARN

You will now start training the model with Seaborn. Perform the following steps to do so:

1. Take a look at the pairplot that you created in the last step of *Exercise 2.03, Analyzing the Boston Housing Dataset*. In particular, look at the scatter plots in the bottom-left corner:

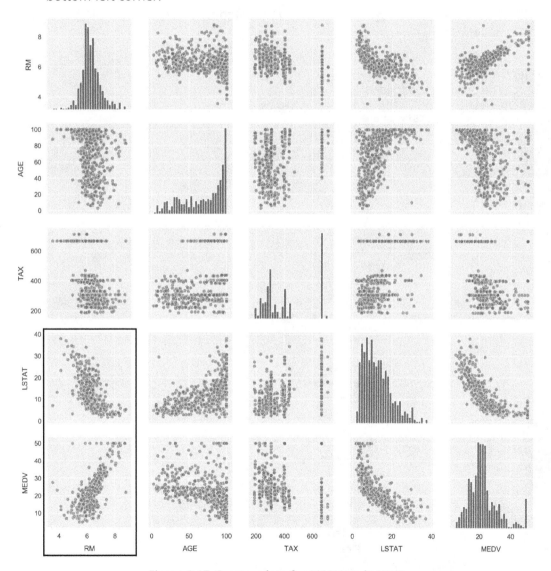

Figure 2.17: Scatter plots for MEDV and LSTAT

Note how the number of **rooms per house (RM)** and the **% of the population that is lower class (LSTAT)** are highly correlated with the **median house value (MDEV)**. Let's pose the following question: how well can we predict MDEV given these variables?

To help answer this, visualize the relationships using seaborn. You will draw the scatter plots along with the line of best fit linear models.

2. To use the **sns.regplot** function (which stands for "regression" plot), pull up that docstring, as follows:

```
sns.regplot?
```

The output is as follows:

```
Signature:
sns.regplot(
    x,
    y,
    data=None,
    x_estimator=None,
    x_bins=None,
    x_ci='ci',
    scatter=True,
    fit_reg=True,
    ci=95,
    n_boot=1000,
    units=None,
    order=1,
    logistic=False,
    lowess=False,
    robust=False,
    logx=False,
    x_partial=None,
    y_partial=None,
    truncate=False,
    dropna=True,
    x_jitter=None,
    y_jitter=None,
    label=None,
    color=None,
    marker='o',
    scatter_kws=None,
    line_kws=None,
    ax=None,
)
Docstring:
Plot data and a linear regression model fit.
```

Figure 2.18: The docstring for sns.regplot

Read about the first few arguments and notice how the function can accept a DataFrame as input.

3. Draw scatter plots for the linear models, as follows:

```
fig, ax = plt.subplots(1, 2)
sns.regplot(x='RM', y='MEDV', data=df, \
            ax=ax[0], scatter_kws={'alpha': 0.4},)
sns.regplot(x='LSTAT', y='MEDV', data=df, \
            ax=ax[1], scatter_kws={'alpha': 0.4},)
plt.savefig('../figures/chapter-2-boston-housing-scatter.png', \
            bbox_inches='tight', dpi=300,)
```

The output is as follows:

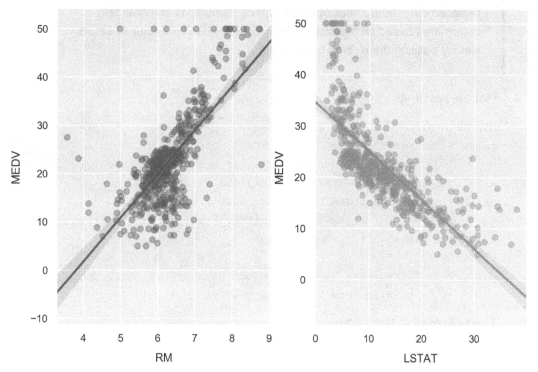

Figure 2.19: Scatter plots with linear regression trends

The line of best fit is calculated by minimizing the ordinary least squares error function, which is something seaborn does automatically when we call the **regplot** function. Also, take note of the shaded areas around the lines, which represent 95% confidence intervals.

> **NOTE**
>
> The 95% confidence intervals that are rendered by the **sns.regplot** function, as seen in the preceding plot, are calculated by taking the standard deviation of data in bins perpendicular to the line of best fit, effectively determining the confidence intervals at each point along the line of best fit. In practice, this involves seaborn bootstrapping the data, a process where new data is created through random sampling with replacement. The number of bootstrapped samples is automatically determined based on the size of the dataset or can be manually set as well by passing the **n_boot** argument.

4. Plot the residuals using seaborn, as follows:

```
fig, ax = plt.subplots(1, 2)
ax[0] = sns.residplot(x='RM', y='MEDV', data=df, \
                      ax=ax[0], scatter_kws={'alpha': 0.4},)
ax[0].set_ylabel('MDEV residuals $(y-\hat{y})$')
ax[1] = sns.residplot(x='LSTAT', y='MEDV', data=df, \
                      ax=ax[1], scatter_kws={'alpha': 0.4},)
ax[1].set_ylabel('')
plt.ylim(-25, 40)
plt.savefig('../figures/chapter-2-boston-housing-residuals.png',\
            bbox_inches='tight', dpi=300,)
```

This will result in the following chart:

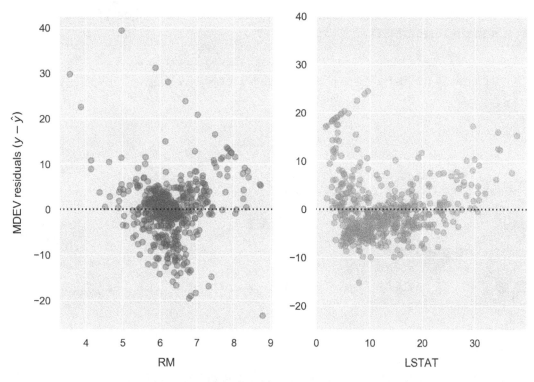

Figure 2.20: Residual charts for linear regression models

Each point on these residual plots is the difference between the observed value (**y**) and the linear model prediction (**ŷ**). Residuals greater than zero are data points that would be underestimated by the model. Likewise, residuals less than zero are data points that would be overestimated by the model.

Patterns in these plots can indicate suboptimal modeling. In each preceding case, you can see diagonally arranged scatter points in the positive region. These are caused by the $50,000 cap on **MEDV**. Both residual charts show data that's largely clustered around 0, which is an indicator of good fit. However, **LSTAT** appears to cluster slightly below 0, indicating that a linear model may not be the best choice. This same fact can be seen by looking at the scatter chart above this, where the line of best fit for **LSTAT** can be seen passing above the bulk of the points.

5. Define a function using scikit-learn that calculates the line of best fit and **mean squared error (MSE)**:

```
from sklearn.linear_model import LinearRegression
from sklearn.metrics import mean_squared_error

def get_mse(df, feature, target='MEDV'):
    # Get x, y to model
    y = df[target].values
    x = df[feature].values.reshape(-1,1)
    print('{} ~ {}'.format(target, feature))
    # Build and fit the model
    lm = LinearRegression()
    lm.fit(x, y)
    msg = ('model: y = {:.3f} + {:.3f}x' \
          .format(lm.intercept_, lm.coef_[0]))
    print(msg)
```

> **NOTE**
>
> The complete code for this step can be found at https://packt.live/2UIzwq8.

In the **get_mse** function, you assign the **y** and **x** variables to the target MDEV and the dependent feature, respectively. These are cast as NumPy arrays by calling the values attribute. The dependent features array is reshaped to the format expected by scikit-learn; this step is only necessary when modeling a one-dimensional feature space. The model is then instantiated and fitted on the data. For linear regression, fitting consists of computing the model parameters using the ordinary least squares method (minimizing the sum of squared errors for each sample). Finally, after determining the parameters, you will predict the target variable and use the results to calculate the MSE. The MSE is simply the sum of squared errors for each data point, where the error is defined as the difference between the observed value and the prediction.

6. Call the **get_mse** function for both **RM** and **LSTAT**, as follows:

```
get_mse(df, 'RM')
get_mse(df, 'LSTAT')
```

The output of calling **RM** and **LSTAT** is as follows:

```
MEDV ~ RM
model: y = -34.671 + 9.102x
mse = 43.60


MEDV ~ LSTAT
model: y = 34.554 + -0.950x
mse = 38.48
```

By comparing the MSE, it transpires that the error is slightly lower for **LSTAT** than for **RM**. Looking back at the scatter plots, however, it appears that we might have even better success using a polynomial model for **LSTAT**. In *Activity 2.01, Building a Third-Order Polynomial Model*, we will test this by computing a third-order polynomial model with scikit-learn.

> **NOTE**
>
> To access the source code for this specific section, please refer to https://packt.live/2N1riF2.
>
> You can also run this example online at https://packt.live/37DzuVK.

Now that we've had an introduction to modeling data with scikit-learn, let's return to our exploration of the Boston Housing dataset. In the next section, we'll introduce some ideas around categorical features in general, and then apply these concepts to our dataset in order to explore the variable relationships in more detail.

USING CATEGORICAL FEATURES FOR SEGMENTATION ANALYSIS

Often, we find datasets where there is a mix of continuous and categorical fields. In such cases, we can learn about our data and find patterns by segmenting the continuous variables with the categorical fields.

As a specific example, imagine you are evaluating the return on investment from an ad campaign. The data you have access to contains measures of some calculated **return on investment** (**ROI**) metric. These values were calculated and recorded daily, and you are analyzing data from the previous year. You have been tasked with finding data-driven insights on ways to improve the ad campaign. Looking at the ROI daily time series, you see a weekly oscillation in the data. Segmenting by day of the week, you find the following ROI distributions (where **0** represents the first day of the week and **6** represents the last):

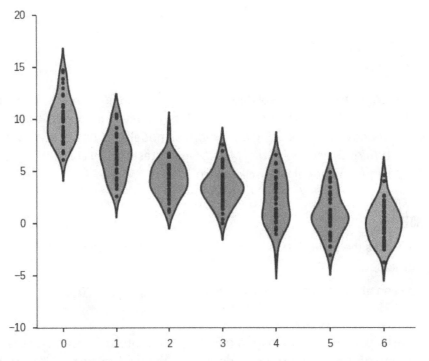

Figure 2.21: A violin plot with ROI on the vertical axis and the day of the week on the horizontal axis

Since we don't have any categorical fields in the Boston Housing dataset we are working with, we'll create one by effectively discretizing a continuous field. In our case, this will involve binning the data into "low", "medium", and "high" categories. It's important to note that we are not simply creating a categorical data field to illustrate the data analysis concepts in this section. As will be seen, doing this can reveal insights from the data that would otherwise be difficult to notice or altogether unavailable.

EXERCISE 2.05: CREATING CATEGORICAL FIELDS FROM CONTINUOUS VARIABLES AND MAKING SEGMENTED VISUALIZATIONS

Before you get started with this exercise, take another look at the final pairplot of *Exercise 2.03*, *Analyzing the Boston Housing Dataset*, where you compared **MEDV**, **LSTAT**, **TAX**, **AGE**, and **RM**:

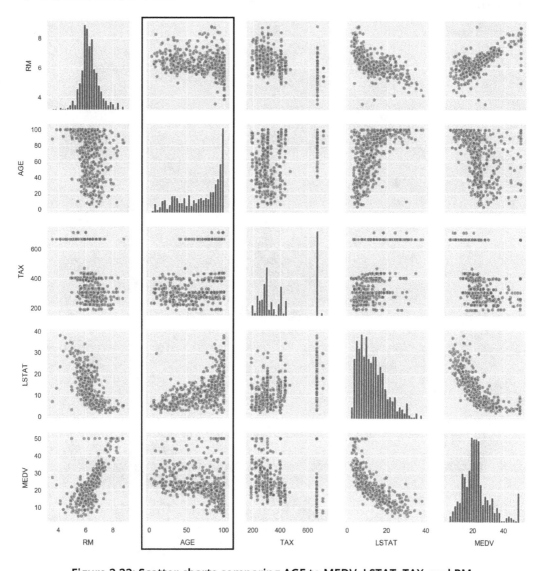

Figure 2.22: Scatter charts comparing AGE to MEDV, LSTAT, TAX, and RM

Take a look at the panels containing **AGE**. As a reminder, this feature is defined as the proportion of owner-occupied units built prior to 1940. We are going to convert this feature into a categorical variable. Once it's been converted, you will be able to replot this figure with each panel segmented by color according to the age category. Perform the following steps to complete this exercise:

1. Plot the **AGE** cumulative distribution, as follows:

```
sns.distplot(df.AGE.values, bins=100, \
            hist_kws={'cumulative': True}, \
            kde_kws={'lw': 0},)
plt.xlabel('AGE')
plt.ylabel('CDF')
plt.axhline(0.33, color='red')
plt.axhline(0.66, color='red')
plt.xlim(0, df.AGE.max())
plt.savefig('../figures/chapter-2-boston-housing-age-cdf.png',\
            bbox_inches='tight', dpi=300,)
```

The output is as follows:

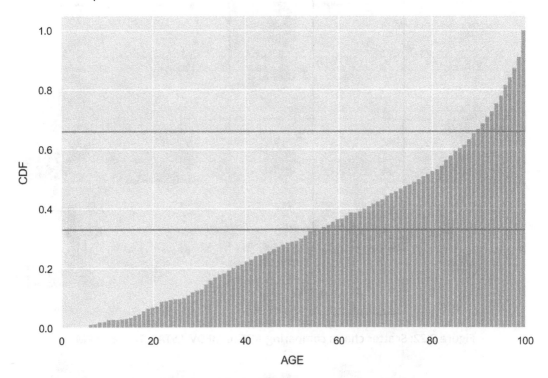

Figure 2.23: Plot for cumulative distribution of AGE

Here, you use **sns.distplot** to plot the distribution and set **hist_kws={'cumulative': True}** in order to calculate and chart the cumulative distribution. In these types of distributions, each bar counts how many values lie in the current x-axis bin and every bin to the left, thereby showing the cumulative amount at that point.

> **NOTE**
>
> Various properties of Seaborn charts can be tuned using **x_kws**, where **x** represents an argument name such as **hist**. Often, these are used to set aesthetic elements of the chart. For example, in the preceding plot, we set **kde_kws={'lw': 0}** in order to bypass plotting the kernel density estimate, which would have been a smooth line corresponding to the density.

Looking at the plot, there are very few samples with a low **AGE** distribution value, whereas there are far more with a very large **AGE**. This is indicated by the steepness of the distribution on the far right-hand side.

The red lines indicate 1/3 and 2/3 points in the cumulative distribution. Looking at the places where our distribution intercepts these horizontal lines, we can see that only about 33% of the samples have the value of **AGE** less than 55, and 33% of the samples have the value of **AGE** greater than 90. In other words, a third of the housing communities have fewer than 55% of their homes built prior to 1940. These would be considered relatively new communities. At the other end of the spectrum, another third of the housing communities have over 90% of their homes built prior to 1940. These would be considered very old. You will use the places where the red horizontal lines intercept the distribution as a guide to split the feature into categories: **Relatively New**, **Relatively Old**, and **Very Old**.

2. Create a new categorical feature and set the segmentation points, as follows:

```
def get_age_category(x):
    if x < 50:
        age = 'Relatively New'
    elif 50 <= x < 85:
        age = 'Relatively Old'
    else:
```

```
        age = 'Very Old'
    return age
```

```
df['AGE_category'] = df.AGE.apply(get_age_category)
```

Here, you are using the very handy pandas **apply** method, which applies a function to a given column or set of columns. The function being applied—in this case, **get_age_category**—should take one argument representing a row of data and return one value for the new column. In this case, the row of data being passed is just a single value, that is, the **AGE** of the sample.

> **NOTE**
>
> The **apply** method is great because it can solve a variety of problems and allows for easily readable code. Often, though, vectorized methods such as **pd.Series.str** can accomplish the same thing much faster. Therefore, it's advised to avoid using it if possible, especially when working with large datasets. We'll see some examples of vectorized methods in *Chapter 3, Preparing Data for Predictive Modeling*.

3. Verify the number of samples you've grouped into each age category, as follows:

```
df.groupby('AGE_category').size()
```

The output is as follows:

```
AGE_category
Relatively New    147
Relatively Old    149
Very Old          210
dtype: int64
```

Looking at the result, you can see that two class sizes are nearly equal and that the **Very Old** group is about 40% larger. You are interested in keeping the classes comparable in size so that each is well represented and it's straightforward to make inferences from the analysis.

> **NOTE**
>
> It may not always be possible to assign samples to classes evenly, and in real-world situations, it's very common to find highly imbalanced classes. In such cases, it's important to keep in mind that it will be difficult to make statistically significant claims with respect to the under-represented class. Predictive analytics with imbalanced classes can be particularly difficult. The following blog post offers an excellent summary regarding the methods for handling imbalanced classes when performing machine learning: https://svds.com/learning-imbalanced-classes/.

Now see how the target variable is distributed when segmented by the new feature, **AGE_category**.

4. Construct a violin plot, as follows:

```
sns.violinplot(x='MEDV', y='AGE_category', data=df, \
               order=['Relatively New', 'Relatively Old', \
                      'Very Old'],)
plt.xlim(0, 55)
plt.savefig('../figures/chapter-2-boston-housing-'\
            'age-medv-violin.png', \
            bbox_inches='tight', dpi=300,)
```

The output is as follows:

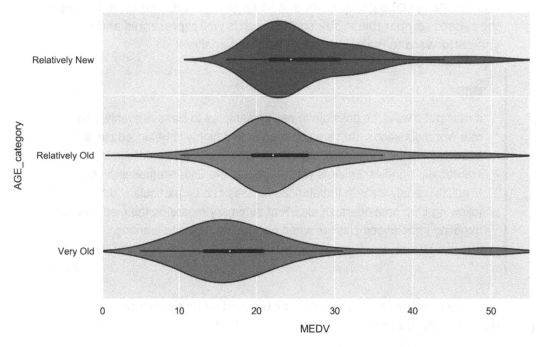

Figure 2.24: Violin plot comparing MEDV distributions by age

The preceding violin plot shows a kernel density estimate of the median house value distribution for each age category. You can compare the distribution of each categorical level with a normal distribution, and you can observe that all three have some degree of skewness. The **Very Old** group contains the lowest median house value records and has a relatively large width, whereas the other groups are more tightly centered on their average. The young group is skewed to the high end, which is evident from the enlarged right half and the position of the white dot in the thick black line within the body of the distribution.

This white dot represents the mean, and the thick black line spans roughly 50% of the population (it fills to the first quantile on either side of the white dot). The thin black line represents boxplot whiskers and spans 95% of the population. This inner visualization can be modified to show the individual data points instead by passing **inner='point' to sns.violinplot()**.

5. Reconstruct the violin plot by adding the **inner='point'** argument to the **sns.violinplot** call. The code for this is as follows:

```
sns.violinplot(x='MEDV', y='AGE_category', data=df, \
               order=['Relatively New', 'Relatively Old', \
                      'Very Old'], \
               inner='point',)
plt.xlim(0, 55)
plt.savefig('../figures/chapter-2-boston-housing-'\
            'age-medv-violin-points.png',\
            bbox_inches='tight', dpi=300,)
```

The output is as follows:

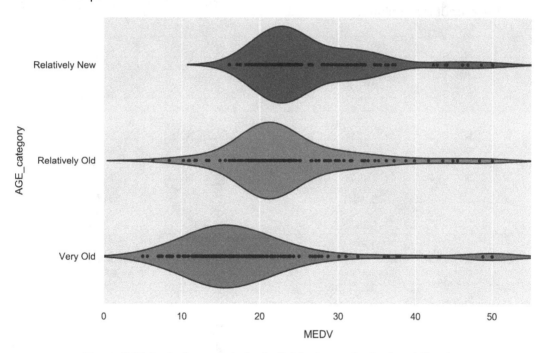

Figure 2.25: Including points for individual samples in the violin plot

It's good to make plots like this for test purposes in order to see how the underlying data connects to the visual. You can see, for example, how there are no median house values lower than roughly $16,000 for the **Relatively New** segment, and therefore, the distribution tail actually contains no data. Due to the small size of our dataset (only about 500 rows), you can see that this is the case for each segment.

6. Reconstruct the **pairplot** visualization from earlier, but now include color labels for each **AGE** category. This is done by simply passing the hue argument, as follows:

```
cols = ['RM', 'AGE', 'TAX', 'LSTAT', 'MEDV', 'AGE_category']
sns.pairplot(df[cols], hue='AGE_category',\
             hue_order=['Relatively New', 'Relatively Old', \
                        'Very Old'], \
             plot_kws={'alpha': 0.5},)
plt.savefig('../figures/chapter-2-boston-housing-'\
            'age-pairplot.png', \
            bbox_inches='tight', dpi=300,)
```

The output is as follows:

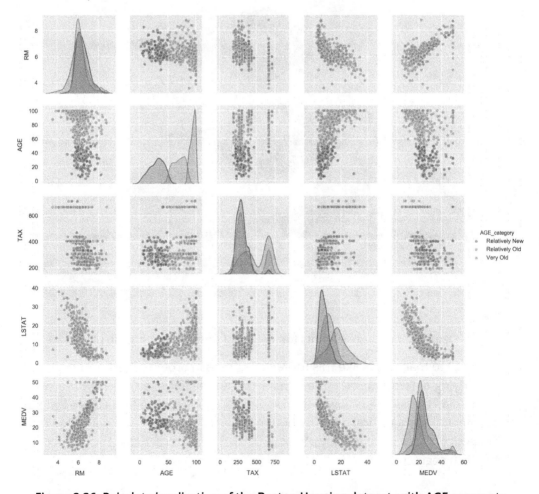

Figure 2.26: Pairplot visualization of the Boston Housing dataset, with AGE segments

Looking at the preceding histograms, you can see that the underlying distributions of each segment appear similar for **RM** and **TAX**. The **LSTAT** distributions, on the other hand, look more distinct. You can focus on them in more detail by using a violin plot.

7. Reconstruct a violin plot to compare the LSTAT distributions for each **AGE_ category** segment, as follows:

```
sns.violinplot(x='LSTAT', y='AGE_category', data=df, \
               order=['Relatively New', 'Relatively Old', \
                      'Very Old'],)
plt.xlim(-5, 40)
plt.savefig('../figures/chapter-2-boston-housing-'\
            'lstat-violin.png',\
            bbox_inches='tight', dpi=300,)
```

The output is as follows:

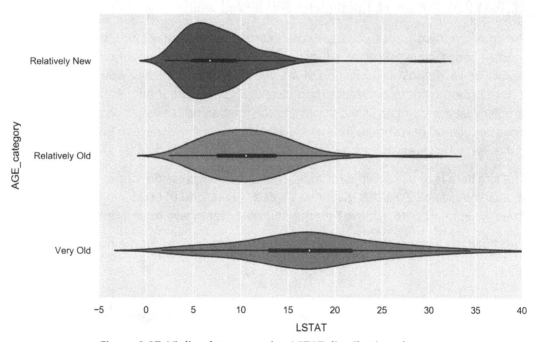

Figure 2.27: Violin plot comparing LSTAT distributions by age

Unlike the **MEDV** violin plot, where each distribution had roughly the same width, here, you can see the width increasing along with **AGE**. Communities with primarily old houses (the **Very Old** segment) contain anywhere from very few to many lower class residents, whereas **Relatively New** communities are much more likely to be predominantly higher class, with over 95% of samples having less lower class percentages than the **Very Old** communities. This makes sense, because **Relatively New** neighborhoods would be more expensive.

> **NOTE**
>
> To access the source code for this specific section, please refer to https://packt.live/2N1riF2.
>
> You can also run this example online at https://packt.live/37DzuVK.

ACTIVITY 2.01: BUILDING A THIRD-ORDER POLYNOMIAL MODEL

Previously in this chapter, you used scikit-learn to create linear models for the median house value (**MEDV**) as a function of **RM** and **LSTAT**, independently. In particular, with **MEDV** as a function of **RM**, you know that modeling would like to build a third-order polynomial model to compare against the linear one. Recall the actual problem we are trying to solve: predicting the median house value, given the lower class population percentage. This model could benefit a prospective Boston House purchaser who cares about how much of their community would be lower class.

The aim here is to use scikit-learn to fit a polynomial regression model to predict the median house value (**MEDV**), given the **LSTAT** values, and to build a model that has a lower MSE. In order to achieve this, the following steps have to be executed:

1. Load the necessary libraries and set up the plot settings for the notebook.

> **NOTE**
>
> While completing this activity, you will need to use many cells in the notebook. Please insert new cells as required.

2. Load the Boston Housing dataset into a pandas DataFrame, as you did earlier in this chapter. Recall that you accessed the data from a built-in scikit-learn dataset. Don't forget to add the target variable column called **MEDV**.

3. Pull out your dependent feature and target variable from **df**. Assign the target to the **y** variable and the feature to the **x** variable. Make sure that **x** that has the proper shape for training scikit-learn models. It should resemble something like this: `[[x1], [x2], [x3], ...]`.

4. Verify that **x** has the proper shape by printing the first three samples.

5. Import the **PolynomialFeatures** class from scikit-learn's preprocessing folder. Then, instantiate it with **degree=3**. After this has been done, display your instantiated object in the notebook.

6. Transform the **LSTAT** feature (assigned to the **x** variable) by running the **fit_transform** method. Assign the newly created polynomial feature set to the **x_poly** variable.

7. Verify what **x_poly** looks like by printing the first few items. They should each have a dimensionality of 4, in order to represent polynomials of degrees 0 through 3.

8. Import the **LinearRegression** class from scikit-learn and train a linear classification model the same way we did while we calculated the MSE. When instantiating the model, set **fit_intercept=False** and assign it the name **clf**.

9. Extract the model coefficients using the **coef_** attribute of **clf**. Use these coefficients to write out the equation of our polynomial model (that is, y = a + bx + cx2 + dx3).

10. Determine the predicted values for each sample and then use these values to calculate the residuals.

11. Print the first 10 residuals. You should see some greater than 0 and some less than 0.

12. Calculate the MSE for your third-order polynomial model.

13. Create a plot of the polynomial model, as a smooth line, overlaying a scatter chart of the samples.

14. Plot the residuals as a scatter chart. Compare this to the linear model residual chart we plotted earlier for this feature. The polynomial model should yield a better fit.

> **NOTE**
>
> The detailed steps, along with the solution to this activity, are presented on page 266.

SUMMARY

In this chapter, we ran an exploratory analysis in a live Jupyter Notebook environment. In doing so, we used visualizations such as scatter plots, histograms, and violin plots to deepen our understanding of the data. We also performed simple predictive modeling, a topic that will be the focus of the following chapters in this book.

In the next chapter, we will discuss how to approach predictive analytics and what things to consider when preparing the data for modeling. We'll use pandas to explore methods of data preprocessing, such as filling missing data, converting from categorical to numeric features, and splitting data into training and testing sets.

3

PREPARING DATA FOR PREDICTIVE MODELING

OVERVIEW

In this chapter, you will learn to plan a machine learning strategy and assess whether or not data is suitable for modeling. You will also perform operations on data in to prepare it so that it can be used to train models. This will include filling missing records, converting from categorical to numeric features, and splitting datasets into training and testing groups. By the end of this chapter, you will be able to process datasets and make them ready for predictive analysis.

INTRODUCTION

Having gone through a fairly involved analysis in the previous chapter, you should now be feeling comfortable using Jupyter Notebooks to work with data. In addition to data exploration and visualization, our analysis included a couple of relatively simple modeling problems, where we trained linear regression models. These *lines of best fit* were very easy to create because only two dimensions were involved and the data was very clean.

As we will see in later chapters, training more advanced models (such as decision trees) can be just as easy because of the simplicity of open source software such as `scikit-learn`. The work involved in preparing data, however, can be significantly more difficult, depending on the details of the relevant datasets.

The quality of training data is very important for creating a model that will generalize well to future samples. For example, errors in your training dataset will cause the model to learn patterns that don't reflect the real-life process behind the data, or act as noise that lowers its accuracy.

Since data preparation is such a big part of the machine learning process, the exercises in this chapter will focus on learning methods that we can use for data processing with Python. We'll continue with our hands-on approach to learning by running through various examples and activities in a Jupyter notebook.

Instead of jumping right into the details of data cleaning, however, we'll introduce machine learning itself and summarize how to approach data science problems in general. Specifically, we will discuss looking at business problems, identifying data that might be useful, and laying out a plan for training predictive models.

MACHINE LEARNING PROCESS

Machine learning (**ML**) lies at the heart of data science. It is an umbrella term for a huge set of algorithms that find and model patterns in data. These algorithms can be broken down into various categories, such as supervised, unsupervised, and reinforcement learning.

In **supervised** problems, we have access to a historical view of labeled records and fit models to predict them—for example, blood test data that's been labeled with the test result. In **unsupervised** problems, there is no such data available, and labels may need to be created using clustering techniques. In later sections, we will break these down in more detail and work with examples of each.

Reinforcement learning is concerned with maximizing a reward function through an iterative process, such as a simulation. Similar to the other types of learning algorithms, there's a wide range of problems that reinforcement learning can be applied to, such as teaching a robot how to walk on a gravel road by adjusting the movement algorithm in order to maximize the distance it can get before falling. This would work roughly as follows:

- Define a set of rules for an agent (the robot) to interact with an environment (gravel road).

- Run a simulation and record the scoring metric (distance moved).

- Adjust the rules based on the scoring metric and repeat this process until no further improvements to the scoring metric can be made.

This type of algorithm is conceptually simple and can be very powerful; however, it's also computationally intense in a large environment with multiple factors to adjust. A famous example of this is AlphaStar, a program that used reinforcement learning to become an expert at playing the computer game StarCraft. It is estimated that AlphaStar has knowledge equivalent to 200 years of human playing time.

> **NOTE**
>
> Reinforcement learning is outside the scope of this book and will not be discussed any further.

The term **learning** in machine learning represents the system's ability to automatically select model parameters and designs from the available data. This process is generally called **fitting** the model to the training data. Since we are concerned with applied data science, you will see how easy this process can be using modern tools from the Python ecosystem.

In contrast, the mathematical and statistical concepts that underpin each modeling method can be very complicated. Well-rounded data scientists should have a solid understanding of these algorithmic details for foundational models such as linear and logistic regression, in addition to anything that's used over the course of a project.

APPROACHING DATA SCIENCE PROBLEMS

It's important to ensure you have a well-structured plan for your data science project before you start the analysis and modeling phases. We'll outline some factors to keep in mind when making this plan, and then go over some technical details regarding preparing data for modeling in the next section.

Since this book is centered around Jupyter Notebooks, we'll start by highlighting how useful they are for the planning phase of a data science project. They offer a very convenient medium for documenting your analysis and modeling plans, for example, by writing rough notes about the data or a list of models we are interested in training. Having these notes in the same place as your proceeding analysis can help others understand what you're doing when they see your work or provide context for you when you look back after leaving it for a while.

A large part of data science involves the use of machine learning to build predictive models. When formulating a plan for predictive modeling, you should start by considering your stakeholder's needs. A perfect model will be useless if it doesn't solve a relevant problem. Planning a strategy around business needs ensures that a successful model will lead to actionable insights.

Although it may be possible in principle to solve many business problems, the ability to deliver the solution will always depend on the availability of the necessary data. Therefore, it's important to consider the business needs in the context of the available data sources. When data is plentiful, this will have little effect, but as the amount of available data becomes smaller, so too does the scope of problems that can be solved.

> **NOTE**
>
> Be careful when it comes to spending too much time on parts of your analysis that are either unrelated to the goals of your project or unlikely to yield useful insights. It's easy to get lost down the rabbit hole while working on a small area of the data or trying to implement a specific feature.
>
> You may spend hours trying to get plots looking just right, when, in the end, nobody except you will end up seeing them. Doing this sort of thing is certainly not without benefit; for example, you will be able to throw together nice-looking charts faster in the future. However, we should not forget to ask ourselves whether the work we're doing is really worth our time with respect to the current project.

These ideas can be formed into a general process for approaching data science problems, such as predictive modeling, which goes as follows:

- **Determine** the business needs by speaking with key stakeholders. Generate ideas for research directions. Seek out a problem where the solution will lead to actionable business decisions.

- **Look** at the available data and try to connect it with the business' needs. Make sure you understand the data fields that are available and the time frames they apply to. Attempt to select a target metric to model (if applicable) and a set of features that can provide insights into business needs.

These steps should be repeated until a realistic plan has taken shape. At that point, you will have identified your model inputs and what outputs are expected.

A lack of available data may cause difficulties or prevent you from identifying a suitable modeling approach. In these cases, keep in mind the possibility of finding new data sources that suit your specific problems, for example, by generating samples through a survey or purchasing third-party datasets.

After identifying a list of problems that might be solved with data modeling, along with the appropriate data sources, we need to lay out a framework for each model. In the next section, we'll look at a series of questions to ask about the data, and the implications of each answer for training models with machine learning.

UNDERSTANDING DATA FROM A MODELING PERSPECTIVE

When attempting to understand a business problem in the context of machine learning, we need to identify whether the problem lends itself to supervised or unsupervised learning. Can the problem be solved by modeling a target variable? If yes, then is this variable available in the dataset? Is it numerical or categorical? The answers to these questions will allow us to identify which modeling algorithms will be relevant to our problem. The following diagram provides an overview of this process:

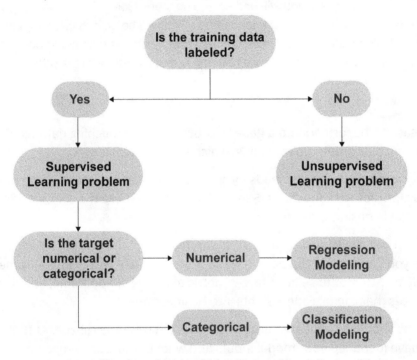

Figure 3.1: A flowchart for identifying types of modeling problems

The preceding flowchart describes the various paths we can follow in order to categorize a dataset for modeling. At the first branch, we are interested in identifying whether a target variable exists. For example, in a weather forecast model, it could be a column recording the amount of rainfall historically. Or perhaps the target variable is a column labeling whether it rained on a given day. The existence of a variable such as this will determine whether it's a supervised or unsupervised learning problem.

Supervised learning can be either a classification or regression problem. In regression, variables are numerical—for example, the amount of rainfall in centimeters. Another situation that could be modeled by regression is movie ratings that range from 1-5 stars.

In classification, the variables are categorical and we predict class labels. The simplest type of classification problem is binary—for example, if trained or not (**yes/no**). An example of multi-class classification is predicting the weather type more generally: sunny, cloudy, rainy, or stormy.

Unsupervised learning problems can be more difficult to solve reliably, and there are fewer approaches for doing so compared to supervised learning. A modeling technique that can be applied to unsupervised problems is cluster analysis, where groups of records are automatically identified based on the distances between the metrics in the feature space. Models of this sort can then assign future records to the nearest cluster.

A good example of unsupervised learning is clustering sets of male or female torso measurements (height, width, arm length, and so on) in order to determine good measurements for each size in a clothing line (S, M, L, XL). The training data should include measurements of people from a random sample of the population so that we have data on a wide range of body types.

When creating the model, it should be limited to four clusters so that clothing measurements for S, M, L, and XL sizes can be selected as the center point of each cluster in the multi-dimensional modeling space. For example, the small size might end up being (x, y, x) = (65 cm, 45 cm, 43 cm), corresponding to the center of a three-dimensional cluster where x = "torso length", y = "torso width", and z = "arm length".

Once you understand the type of problem to model, there are other factors to consider, as follows:

- The size of data, in terms of the width (number of columns) and height (number of rows).

- Algorithm selection, depending on the details of the features. Some features apply themselves to certain algorithms over others. For example, sparse datasets (lots of zeros) are better modeled with **Lasso regression**, rather than standard ordinary least squares regression.

> **NOTE**
>
> The Lasso algorithm regularizes the model by adding a penalty to the cost function that depends on the input sample values. This has the effect of dropping weight parameters to zero, hence hiding certain features from the model—that is, dimensionality reduction.

- Generally, larger datasets perform better in terms of accuracy, compared to smaller datasets of the same underlying process.

- Training models on large amounts of data can be time-consuming. Sometimes, it's possible to save time on large datasets by using dimensionality reduction techniques to reduce the number of features.

- A set of various different models should be trained and compared for a given problem. Combining high-accuracy models using ensemble averaging techniques can increase overall accuracy, resulting in models that perform better on unseen data.

These considerations assume that the modeling data exists in a single table. However, the existence of multiple interconnected datasets will require additional high-level model design choices.

For example, consider the following situation, where we have two data sources:

- A data feed (for example, a table) with the AAPL stock closing prices on a daily timescale

- A data feed (for example, a table) with iPhone sales data on a monthly timescale

The difficulty here is dealing with the different timescales of each dataset.

One approach is to merge them, by adding the monthly sales data to each sample in the daily timescale table or by grouping the daily data by month.

Another idea would be to build two models, one for each dataset, and use a combination of the results from each in the final ensemble prediction model.

These early-stage modeling decisions should be considered carefully during the preparation phase in order to avoid going down the wrong path. Oftentimes, however, design choices will arise where the best option is not obvious in the planning phase. In cases such as these (after considering trade-offs such as accuracy, computational complexity, and interpretability), the best option should be determined by training multiple models and comparing them using the relevant scoring metrics.

Now that we've discussed high-level concepts around data modeling, we'll focus our attention on preparing data for predictive modeling.

PREPARING DATA FOR MODELING

In order to build models that will perform well on unseen data (that is, that work well in production), we must train them on carefully processed data. Like the saying *"you are what you eat"*, the model's performance is a direct reflection of the data it's trained on. Of course, its performance will also depend on your ability to avoid overfitting or underfitting models, as will be discussed in *Chapter 5*, *Model Validation and Optimization*, and *Chapter 6*, *Web Scraping with Jupyter Notebooks*.

The aspect of data preprocessing for machine learning that usually takes the longest is cleaning up messy data. Some estimates suggest that data scientists spend around two thirds of their work time cleaning and organizing datasets. This includes tasks such as the following:

- Merging datasets on common fields to bring all the data into a single table

- Feature engineering to improve the quality of the data

- Removing or filling incorrect or missing values, for example, by replacing missing data with the mean or median of the existing values for that field

- Dropping duplicate records

- Building the training datasets by standardizing or normalizing the required data, applying feature transformations, and applying train-test splits

Now, let's learn the basics of how these tasks can be done using Jupyter Notebooks and pandas.

In the following exercise, we will encounter missing (NaN) values for the first time in this book. So, let's discuss how these work in Python.

You can define an NaN variable by doing, for example, `a = float('nan')`. This is the data structure that pandas uses to express missing values.

One of the tricky things about working with NaNs is testing for equality. You cannot simply use standard comparison methods as they will behave unexpectedly. Instead, it's best to make comparisons with a high-level function from a library such as pandas or NumPy. This is illustrated with the following code:

```
[1]:  a = float('nan')

[2]:  bool(a)

[2]:  True

[3]:  a == float('nan')

[3]:  False

[4]:  a is float('nan')

[4]:  False

[5]:  import numpy as np
      np.isnan(a)

[5]:  True

[6]:  import pandas as pd
      pd.isna(a)

[6]:  True
```

Figure 3.2: Working with missing values in Python

Some of these results may seem counterintuitive. There is logic behind this behavior, however, and for a deeper understanding of the fundamental reasons for standard comparisons returning **False**, check out this excellent StackOverflow answer: https://stackoverflow.com/a/1573715/3511819.

EXERCISE 3.01: DATA CLEANING FOR MACHINE LEARNING WITH PANDAS

For the remainder of this chapter, we'll explore how data is transformed with the pandas library by working on a very simple demonstration dataset.

In *Activity 3.01, Preparing to Train a Predictive Model for Employee Retention*, we'll apply the techniques from this exercise *in practice* and clean up a significantly larger dataset.

For now, though, we are going to focus on some basics of the pandas library. We'll learn how to merge tables, handle duplicated records, and deal with missing values. Perform the following steps to complete this exercise:

1. If you haven't done so already, start up one of the following platforms for running Jupyter Notebooks:

   JupyterLab (run **jupyter lab**)

   Jupyter Notebook (run **jupyter notebook**)

   Then, open up the platform you chose in your web browser by copying and pasting the URL, as prompted in the Terminal:

2. Load the required libraries and set up your plot settings for the notebook, as follows:

```
import pandas as pd
import numpy as np
import datetime
import time
import os

import matplotlib.pyplot as plt
%matplotlib inline
import seaborn as sns

%config InlineBackend.figure_format='retina'
sns.set() # Revert to matplotlib defaults
plt.rcParams['figure.figsize'] = (9, 6)
plt.rcParams['axes.labelpad'] = 10
sns.set_style("darkgrid")
```

3. Now, we are going to start by showing off some basic tools from pandas and scikit-learn. Load the watermark magic extension as shown here:

```
%load_ext watermark
%watermark -d -v -m -p \
requests,numpy,pandas,matplotlib,seaborn,sklearn
```

4. Generate two sample datasets and merge them. Display the docstring for the **merge** function, as follows:

```
pd.merge?
```

The output is as follows:

```
Signature:
pd.merge(
    left,
    right,
    how='inner',
    on=None,
    left_on=None,
    right_on=None,
    left_index=False,
    right_index=False,
    sort=False,
    suffixes=('_x', '_y'),
    copy=True,
    indicator=False,
    validate=None,
)
Docstring:
Merge DataFrame or named Series objects with a database-style join.

The join is done on columns or indexes. If joining columns on
columns, the DataFrame indexes *will be ignored*. Otherwise if joining indexes
on indexes or indexes on a column or columns, the index will be passed on.

Parameters
----------
left : DataFrame
right : DataFrame or named Series
    Object to merge with.
how : {'left', 'right', 'outer', 'inner'}, default 'inner'
    Type of merge to be performed.
```

Figure 3.3: The docstring for pd.merge

As you can see, the function accepts a **left** and **right** DataFrame to merge. You can specify one or more columns to group on, as well as how they are grouped—that is, to use the left, right, outer, or inner sets of values. Start this exercise by experimenting with this function so that we can learn how to use it.

5. Now, define our example datasets:

```
df_1 = pd.DataFrame({'product': ['red shirt', 'red shirt', \
                                 'red shirt', 'white dress',], \
                'price': [49.33, 49.33, 32.49, 199.99,],})
df_2 = pd.DataFrame({'product': ['red shirt', 'blue pants', \
                                 'white tuxedo', 'white dress',], \
                'in_stock': [True, True, False, False,],})
```

Here, you will build two simple DataFrames from scratch. As can be seen, they contain a common key (the product column) that has some shared entries between the two tables.

6. Display the first DataFrame as follows:

```
df_1
```

The output is as follows:

	product	price
0	red shirt	49.33
1	red shirt	49.33
2	red shirt	32.49
3	white dress	199.99

Figure 3.4: The first product's DataFrame

7. Display the second DataFrame as follows:

```
df_2
```

The output is as follows:

	product	in_stock
0	red shirt	True
1	blue pants	True
2	white tuxedo	False
3	white dress	False

Figure 3.5: The second product's DataFrame

8. Now, use the **merge** function from pandas to perform an inner merge and display the result, as follows:

```
df = pd.merge(left=df_1, right=df_2, \
          on='product', how='inner')
df
```

The following screenshot shows the output of the inner **merge** function:

	product	price	in_stock
0	red shirt	49.33	True
1	red shirt	49.33	True
2	red shirt	32.49	True
3	white dress	199.99	False

Figure 3.6: Inner merge on products

Note how only the shared items (that is, red shirt and white dress) are included. To include all entries from both tables, you can do an outer merge instead, which you will do in the next step.

9. Perform an outer merge and display the result, as follows:

```
df = pd.merge(left=df_1, right=df_2, \
          on='product', how='outer')
df
```

The following screenshot shows the output of the outer **merge** function:

	product	price	in_stock
0	red shirt	49.33	True
1	red shirt	49.33	True
2	red shirt	32.49	True
3	white dress	199.99	False
4	blue pants	NaN	True
5	white tuxedo	NaN	False

Figure 3.7: Outer merge on products

This returns all data from each table where missing values have been labeled with **NaN**.

10. You may have noticed that our most recently merged table has duplicated data in the first few rows. To handle this, return a version of the DataFrame with no duplicate rows by using the following code:

```
df.drop_duplicates()
```

The output is as follows:

	product	price	in_stock
0	red shirt	49.33	True
2	red shirt	32.49	True
3	white dress	199.99	False
4	blue pants	NaN	True
5	white tuxedo	NaN	False

Figure 3.8: Merged products' table after dropping duplicates

This is the easiest way to drop duplicate rows. To apply these changes to **df**, you can either set the **inplace=True** argument or run **df = df.drop_duplicated()**.

Now consider another method that uses masking to select or drop duplicate rows.

11. Print the **True/False** series in order to mask duplicate rows, as follows:

```
df.duplicated()
```

The output is as follows:

```
0    False
1     True
2    False
3    False
4    False
5    False
dtype: bool
```

12. Sum the result to determine how many rows have been duplicated. Use the following code to do so:

```
df.duplicated().sum()
```

This will return **1**.

13. Display this duplicated row as follows:

```
df[df.duplicated()]
```

The output is as follows:

	product	price	in_stock
1	red shirt	49.33	True

Figure 3.9: Displaying a duplicated row

14. In the previous step, you used a mask to filter the DataFrame on duplicated rows. By using the tilde symbol, ~, you can use the same technique to show the other set of rows from the DataFrame. Use the following code to do so:

```
df[~(df.duplicated())]
```

The output showing the use of ~ before the **duplicated** function is as follows:

	product	price	in_stock
0	red shirt	49.33	True
2	red shirt	32.49	True
3	white dress	199.99	False
4	blue pants	NaN	True
5	white tuxedo	NaN	False

Figure 3.10: Displaying non-duplicated rows

15. Using this masking technique allows for more granular control over how duplicates can be treated. For example, eliminate duplicate products using the following code:

```
mask = ~(df['product'].duplicated())
df[mask]
```

The output of the preceding command for removing duplicates is as follows:

	product	price	in_stock
0	red shirt	49.33	True
3	white dress	199.99	False
4	blue pants	NaN	True
5	white tuxedo	NaN	False

Figure 3.11: Merged products' table after dropping duplicate products

By running this line, you do the following:

Create a mask (a **True/False** series) for the product row, where duplicates are marked with **True**.

Take the opposite of that mask, using the tilde symbol (~), so that duplicates are marked with **False** and everything else is **True**.

Use that mask to filter out the **False** rows of **df**, which correspond to the duplicated products.

As expected, we can now see that only the first red shirt row remains, as the duplicate product rows have been removed.

16. At this point, you have tested various methods for dropping duplicates, but have not actually changed **df**. In order to proceed with this exercise, replace **df** with a deduplicated version of itself. This can be done as follows:

```
df.drop_duplicates(inplace=True)
df
```

The following outputs show **df** after dropping duplicates:

	product	price	in_stock
0	red shirt	49.33	True
2	red shirt	32.49	True
3	white dress	199.99	False
4	blue pants	NaN	True
5	white tuxedo	NaN	False

Figure 3.12: Merged products' table after dropping duplicate rows

As can be seen, the duplicate row has now been permanently dropped from **df**.

The next aspect of preprocessing you will learn about is handling missing data. This is important because many models cannot be trained on incomplete records.

17. One option is to drop the rows, which might be a good idea if your NaN records are missing the majority of their values. Use the following code to do so:

```
df.dropna()
```

Similar to most of the commands you ran earlier for dropping duplicates, calling **df.dropna()** will not alter the DataFrame. Instead, it returns a new DataFrame that is rendered for you in the notebook, below the cell. As expected, the rows with missing data for the "blue pants" and "white tuxedo" products are not present here:

	product	price	in_stock
0	red shirt	49.33	True
2	red shirt	32.49	True
3	white dress	199.99	False

Figure 3.13: Merged products' table after dropping rows with missing values

18. If most of the values are missing for a feature, it may be best to drop that column entirely. This can be done with the same method as before, by setting the axes argument to **1**, in order to specify columns instead of rows. This can be done as follows:

```
df.dropna(axis=1)
```

The output is as follows:

	product	in_stock
0	red shirt	True
2	red shirt	True
3	white dress	False
4	blue pants	True
5	white tuxedo	False

Figure 3.14: Merged products' table after dropping columns with missing values

Simply dropping the **NaN** values, just like we did for the last two examples, is usually not the best option. This is because this causes us to lose training data that may be valuable.

Instead of dropping them, we can consider filling the missing entries. Pandas offers a method for filling in **NaN** entries in a variety of different ways, some of which we'll illustrate now.

19. Print the docstring for the pandas **NaN-fill** method, as follows:

```
df.fillna?
```

The output is as follows:

```
Signature:
df.fillna(
    value=None,
    method=None,
    axis=None,
    inplace=False,
    limit=None,
    downcast=None,
    **kwargs,
)
Docstring:
Fill NA/NaN values using the specified method.

Parameters
----------
value : scalar, dict, Series, or DataFrame
    Value to use to fill holes (e.g. 0), alternately a
    dict/Series/DataFrame of values specifying which value to use for
    each index (for a Series) or column (for a DataFrame).  Values not
    in the dict/Series/DataFrame will not be filled. This value cannot
    be a list.
method : {'backfill', 'bfill', 'pad', 'ffill', None}, default None
    Method to use for filling holes in reindexed Series
    pad / ffill: propagate last valid observation forward to next valid
```

Figure 3.15: The docstring for pd.DataFrame.fillna

Note the options for the **value** parameter; this could be, for example, a single value or a dictionary/series type map based on the index. Alternatively, you can leave the value as **None** and pass a `fill` method instead. We'll see examples of each later on.

20. Print the DataFrame as follows:

```
df
```

The output is as follows:

	product	price	in_stock
0	red shirt	49.33	True
2	red shirt	32.49	True
3	white dress	199.99	False
4	blue pants	NaN	True
5	white tuxedo	NaN	False

Figure 3.16: The merged products' table

21. One strategy for filling missing data is to use the average value for that field. Do this for the price column by running the following code:

```
fill_value = df.price.mean()
df.fillna(value=fill_value)
```

The output is as follows:

	product	price	in_stock
0	red shirt	49.330000	True
2	red shirt	32.490000	True
3	white dress	199.990000	False
4	blue pants	93.936667	True
5	white tuxedo	93.936667	False

Figure 3.17: Merged products' table after filling in the missing prices with the mean

22. In some cases, you may want to fill the missing values using the previous non-null record. Note that this strategy will depend on the sorting order of the DataFrame.

Do this for the **price** column as follows:

```
df.fillna(method='pad')
```

The output is as follows:

	product	price	in_stock
0	red shirt	49.33	True
2	red shirt	32.49	True
3	white dress	199.99	False
4	blue pants	199.99	True
5	white tuxedo	199.99	False

Figure 3.18: Merged products' table after filling in the missing prices with the "pad" method

Notice how the **white dress** price was used to pad the missing values below it.

23. Now that you have looked at a couple of the ways in which pandas enables you to fill missing data, go with the mean value method. Permanently replace the missing data in **df**, as follows:

```
df = df.fillna(value=df.price.mean())
```

This concludes the first of our exercises on preparing data for machine learning. We have learned how to merge tables, identify issues such as duplicate rows and missing values, and then solved those issues using Pandas.

> **NOTE**
>
> To access the source code for this specific section, please refer to https://packt.live/2YzXB3j.
>
> You can also run this example online at https://packt.live/2Y3vvi4.

EXERCISE 3.02: PREPARING DATA FOR MACHINE LEARNING WITH PANDAS

In this exercise, we will continue working with the simple table from earlier and finish preparing it so that it can be used to train a machine learning algorithm. We won't actually try to train any models on such a small dataset, though! When we start training models in *Chapter 4, Training Classification Models*, we'll be using a much more realistic table that has roughly 15k rows.

We'll start this process by encoding the class labels for the categorical data, and then complete this exercise by performing train/test splits on the data. Perform the following steps to complete this exercise:

> **NOTE**
>
> This exercise is built on top of *Exercise 3.01, Data Cleaning for Machine Learning with Pandas*, and should be executed in the same notebook.

1. Before learning about label encoding, create a new field that represents the average product ratings by running the following code:

```
ratings = ['low', 'medium', 'high']
np.random.seed(2)
df['rating'] = np.random.choice(ratings, len(df))
df
```

The output is as follows:

	product	price	in_stock	rating
0	red shirt	49.330000	True	low
2	red shirt	32.490000	True	medium
3	white dress	199.990000	False	low
4	blue pants	93.936667	True	high
5	white tuxedo	93.936667	False	high

Figure 3.19: The table after adding a new column for rating

2. Now, encode all non-numeric data types present in the table. The simplest column to handle is the Boolean list called **in_stock**. This can easily be mapped to binary numeric values (that is, **0** and **1**). Use the following code to do so:

```
df.in_stock = df.in_stock.map({False: 0, True: 1})
df
```

The output is as follows:

	product	price	in_stock	rating
0	red shirt	49.330000	1	low
2	red shirt	32.490000	1	medium
3	white dress	199.990000	0	low
4	blue pants	93.936667	1	high
5	white tuxedo	93.936667	0	high

Figure 3.20: The table after converting the in_stock column into numeric form

3. Another option for encoding feature labels is to use scikit-learn's **LabelEncoder** library. This gives you a high-level abstraction to perform encoding using scikit-learn's API. Test this using the following code:

```
from sklearn.preprocessing import LabelEncoder
rating_encoder = LabelEncoder()
df.rating = rating_encoder.fit_transform(df.rating)
df
```

The output is as follows:

	product	price	in_stock	rating
0	red shirt	49.330000	1	1
2	red shirt	32.490000	1	2
3	white dress	199.990000	0	1
4	blue pants	93.936667	1	0
5	white tuxedo	93.936667	0	0

Figure 3.21: The table after converting rating into a numeric form

This might bring to mind the preprocessing you did in the previous chapter, when building the polynomial model. Here, instantiate a label encoder and then "train" it and "transform" your data using the **fit_transform** method:

```
rating_encoder.inverse_transform(df.rating)
```

This displays the following output:

```
array(['low', 'medium', 'low', 'high', 'high'], dtype=object)
```

4. One benefit of using scikit-learn's **LabelEncoder** is that you can convert the feature values back into their original form using the **inverse_transform** function. Do this by running the following code. Study the output to convince yourself that they are the proper labels:

```
df.rating = rating_encoder.inverse_transform(df.rating)
df
```

The output is as follows:

	product	price	in_stock	rating
0	red shirt	49.330000	1	low
2	red shirt	32.490000	1	medium
3	white dress	199.990000	0	low
4	blue pants	93.936667	1	high
5	white tuxedo	93.936667	0	high

Figure 3.22: The table after performing an inverse transform on the rating column

You may notice a problem here. We are working with a so-called **ordinal feature**, where there's an inherent order to the labels. In this case, we should expect a rating of **low** to be encoded with a **0**, and a rating of **high** to be encoded with a **2**. However, this is not the result we can see. In order to achieve proper ordinal label encoding, we should build a mapping dictionary ourselves.

5. Encode the ordinal labels properly, as follows:

```
ordinal_map = {rating: index \
            for index, rating in enumerate(['low', \
                                            'medium', \
                                            'high'])}
print(ordinal_map)
df.rating = df.rating.map(ordinal_map)
df
```

The output is as follows:

```
{'low': 0, 'medium': 1, 'high': 2}
```

	product	price	in_stock	rating
0	red shirt	49.330000	1	0
2	red shirt	32.490000	1	1
3	white dress	199.990000	0	0
4	blue pants	93.936667	1	2
5	white tuxedo	93.936667	0	2

Figure 3.23: The table after mapping the ordinal values for rating

Here, we create the **ordinal_map** mapping dictionary. This is done using dictionary comprehension and enumeration, but looking at the result, we can see that it could just as easily be defined manually instead. Of course, this is only the case for our example because there are very few unique values in the **rating** field. We might expect the number of unique values to be very large, in which case the dictionary comprehension method would be a lot more useful.

As we did earlier for the **in_stock** column, we applied dictionary mapping to the feature. Looking at the result, we can see that rating makes more sense than before, where **low** is labeled with **0**, **medium** with **1**, and **high** with **2**.

Now that we've discussed ordinal features, let's touch on another type called **nominal features**. These are fields with no inherent order, and in our case, we can see that a product is a perfect example.

Most scikit-learn models can be trained on data like this, where we have strings instead of integer-encoded labels. In this situation, the necessary conversions are done under the hood. However, this may not be the case for all models in scikit-learn, or other machine learning and deep learning libraries. Therefore, it's good practice to encode these ourselves during the preprocessing stage.

6. A commonly used technique to convert class labels from strings into numerical values is called one-hot encoding. This splits the distinct classes into separate features, and can be accomplished elegantly with **pd.get_dummies()**. Do this as follows:

```
df = pd.get_dummies(df)
df
```

The output is as follows:

	price	in_stock	rating	product_blue pants	product_red shirt	product_white dress	product_white tuxedo
0	49.330000	1	0	0	1	0	0
2	32.490000	1	1	0	1	0	0
3	199.990000	0	0	0	0	1	0
4	93.936667	1	2	1	0	0	0
5	93.936667	0	2	0	0	0	1

Figure 3.24: The result of calling get_dummies on the table

Here, we can see the result of one-hot encoding: the **product** column has been split into four, one for each unique value. Within each column, we find either a **1** or **0**, representing whether that row contains a particular value or product.

> **NOTE**
>
> By one-hot encoding the variables like we did here, where the four unique values were broken out into four new columns, we have put ourselves at risk of introducing **multicollinearity** into our dataset. This occurs when one or more variables in a dataset can be described by a linear combination of others. Having a linear dependency such as this can lead to issues when modeling, so it's best to take efforts to avoid it.
>
> One method that fixes the multicollinearity issue introduced by one-hot encoding is to drop one of the resulting columns. This can be done with **pd.get_dummies(df, drop_first=True)**. There are alternate ways of handling multicollinearity that can also be considered; however, this topic is beyond the scope of this book and we will not worry about its consequences here.

Moving on and ignoring any data scaling (which should usually be done), the final step is to split the data into training and test sets so that we can use them for machine learning. This can be done using scikit-learn's **train_test_split**. Let's assume we are going to try to predict whether an item is in stock, given the other feature values.

7. Split the data into training and test sets by running the cell containing the following code:

```
features = ['price', 'rating','product_blue pants', \
            'product_red shirt','product_white dress', \
            'product_white tuxedo',]
X = df[features].values

target = 'in_stock'
y = df[target].values

from sklearn.model_selection import train_test_split
X_train, X_test, \
y_train, y_test = (train_test_split(X, y, \
                        test_size=0.3))
```

Here, you select subsets of the data and feed them into the **train_test_split** function. This function has four outputs, which are unpacked into the training and testing splits for the features (**X**) and the target (**y**).

8. Print the shapes of each result that was unpacked from the **train_test_split** function by running the following code:

```
print('Data Shapes')
print('--------------')
print('X_train', X_train.shape)
print('X_test ', X_test.shape)
print('y_train', y_train.shape)
print('y_test ', y_test.shape)
```

The output is as follows:

```
Data Shapes
--------------
X_train (3, 6)
X_test  (2, 6)
y_train (3,)
y_test  (2,)
```

Observe the shape of the output data, where the test set should have roughly 30% of the records and the training set should have roughly 70%. Since we only have five records here, scikit-learn splits the data into two testing records and three training records. When preparing a proper dataset, however, the splits will be quite close to the specified threshold.

> **NOTE**
>
> When we call the values attribute in the preceding code, we are converting the pandas series (that is, the DataFrame column) into a NumPy array. This is good practice because it strips out unnecessary information from the series object, such as the index and name.

This concludes the training exercise on cleaning data so that it can be used in machine learning applications.

> **NOTE**
>
> To access the source code for this specific section, please refer to https://packt.live/2YzXB3j.
>
> You can also run this example online at https://packt.live/2Y3vvi4.

Did you notice how effective our Jupyter notebook was for testing various methods of transforming data? Be sure to keep lab-style notebooks such as this so you can look back on your work later and understand the data cleaning decisions that were made.

This notebook could also be used to re-process an updated version of the data and, should we wish to make any changes to the processing, these can easily be tested in the notebook by altering the appropriate cells. The best way to achieve this would probably be to copy the notebook over to a new file so that we can always keep a copy of the original analysis for reference.

In the next section, we'll apply the concepts from this exercise to a large dataset as we prepare it so that it can be used to train predictive models.

INTRODUCING THE HUMAN RESOURCE ANALYTICS DATASET

Having learned about basic data cleaning concepts and seen them implemented with pandas and scikit-learn, we'll put what we've learned into practice on a diverse dataset that has real-world context. In the following chapters, we'll model this dataset with a variety of machine learning techniques, so let's take some time to get familiar with it now. Let's imagine the following situation:

Suppose you are hired to do freelance work for a company who wants to find insights into why their employees are leaving. They have compiled a set of data they think will be helpful in this respect. It includes details of employee satisfaction levels, evaluations, time spent at work, department, and salary.

The company shares their data with you by sending you a file called **hr_data.csv** and asks you what you think can be done to help stop employees from leaving.

Our aim is to apply the concepts we've discussed thus far to a real-life problem. In particular, we seek to do the following:

- Determine a plan for using data modeling to provide impactful business insights, given the available data.

- Prepare the dataset for use in machine learning models.

Having identified that the relevant business problem is employee turnover, we would generally want to have a big-picture discussion around why that might be happening and what data would be helpful for generating solutions.

For the purposes of this example, let's start by looking at the dataset provided by the company. We need to understand the basic properties of the dataset, such as the number of samples and column descriptions. The basic properties of this dataset are as follows:

- Rows: 15,000 samples (rows) representing previous or current employees

- Columns: The columns and data types are as follows:

  left (if employee has left the company or not [bool variable])

  satisfaction_level, **last_evaluation**, **average_montly_hours**, **time_spend_company**, **number_project** (employee metrics [float variables])

  work_accident, **promotion_last_5years**, **is_smoker** (employee metrics [bool variables])

  department, **salary** (employee metrics [string variables])

After reviewing this information, we can see the existence of a large dataset (15k rows) with the relevant signals for employee turnover. Namely, these are metrics such as satisfaction level, time spent working, and department, which may reveal interesting patterns that shed light on the reasons for employees leaving the company.

This dataset is labelled with a **yes/no** (Boolean) variable, indicating whether an employee has turned over (**left**). By referencing the data flowchart earlier in this chapter, we can see that labeled data such as this applies itself well to supervised learning models. In other words, we could try to predict the **yes/no** variable **left** column, given the other variables. However, would this be helpful for the business?

Let's think again about the business' needs: the company wants to reduce the number of employees who leave. If we were able to predict how likely an employee is to quit, the business could selectively target those employees for special treatment; for example, their salary could be raised or their number of projects could be reduced. Furthermore, the impact of these specific salary or project changes could be estimated using our model.

Given this business-oriented discussion, it seems like we have reached a decision on how to approach modeling: our goal will be to train a model that can predict whether an employee has left.

> ### NOTE
>
> As we discussed previously, we'll be using the Human Resource Analytics dataset, which was originally found in Kaggle's open source repository. The link to this book's version of the dataset can be found on GitHub: https://packt.live/3hEBQIy.

This data has been simulated, meaning the samples are artificially generated and do not represent real people. We'll ignore this fact and model the data as if it were generated by real-life processes.

Note that there's a small difference between the dataset we're using in this book and the original Kaggle version. Our Human Resource Analytics dataset contains some missing values (NaNs) so that we can illustrate data cleaning techniques. We have also added a column of data called `is_smoker` for the same purposes.

We will do this by modeling the `left` target variable as a function of the other features (columns) listed previously.

Now that we have an idea of what type of model we wish to create, we need to become familiar with the details of the dataset and prepare it for modeling. These two points will be the focus of the following activity, which rounds off this chapter.

ACTIVITY 3.01: PREPARING TO TRAIN A PREDICTIVE MODEL FOR EMPLOYEE RETENTION

In this activity, you'll start exploring the Human Resource Analytics dataset and prepare it for modeling. You will load the dataset into a Jupyter notebook and look at the data types, distribution, and missing values in each column. Then, you'll clean the data by identifying and fixing issues that would lead to issues when modeling.

1. If you haven't done so already, start up one of the following platforms for running Jupyter Notebooks:

    JupyterLab (run **`jupyter lab`**)

    Jupyter Notebook (run **`jupyter notebook`**)

    Then, open up the platform you have chosen in your web browser by copy and pasting the URL, as prompted in the Terminal.

2. The first step is to load the dataset into the notebook. Before doing this, use bash to print the first 10 or so rows of the table (the **head**) in your notebook. If you're unable to use bash with your Jupyter environment, then do this with Python code.

3. Load the table with pandas and assign it to the **df** variable.

4. Print the table columns, followed by the first few rows (the **head**) and the last few rows (the **tail**).

5. Open the CSV file with Python (using **open** instead of pandas) and count the number of rows in the table.

6. Calculate the length of **df** and compare with the preceding result. Do the numbers match? If not, why not?

7. Check how the **left** variable is distributed. How many **yes** values and how many **no** values are there? Are there any missing values?

8. Print the data type of each feature.

9. Plot histograms/bar charts in order to visualize each feature distribution.

10. Check how many NaN values there are in each column. For the features with missing data, think about what you would do to fix them. This could include filling the missing data, dropping the feature, or dropping samples.

11. Drop the **is_smoker** column from **df**.

12. Fill the NaN values in the **time_spend_company** column with the median value.

13. Make a boxplot of **average_montly_hours** segmented by **number_project**. How can this chart help you fill the missing **average_montly_hours** data?

14. Calculate the mean **average_montly_hours** data of each segment from the preceding chart.

15. Fill the NaN values in **average_montly_hours** using the means calculated previously for each segment. This can be done by passing a pandas series object to **df.fillna**. This series should have the same shape as **df**, with the appropriate values to fill for each NaN entry.

16. Confirm that **df** has no more NaN values.

17. Now that your data has been cleaned, the last step is to transform non-numeric fields into integer representations. Boolean fields should be mapped to **0** and **1**, while categorical fields should be one-hot encoded.

18. Print the columns of our transformed dataset. There should be new columns visible after one-hot encoding.

19. Save our preprocessed data in a CSV file.

> **NOTE**
>
> The detailed steps for this activity, along with the solutions and additional commentary, are presented on page 273.

After completing this activity, the Human Resource Analytics dataset is now ready to model.

Before moving on, let's briefly pause here to note how well-suited Jupyter notebooks are for performing this initial data analysis and cleanup.

Imagine, for example, that we left this project in its current state for a few months. Upon returning to it, we would probably not remember what exactly was going on when we left it. However, by referring back to this notebook, we would be able to retrace our steps and quickly recall what we previously learned about the data.

Furthermore, we could use a fresh dataset and rerun the notebook to prepare the new set of data for use in our machine learning algorithms. Recall that in this situation, it would be best to make a copy of the notebook first, so as not to lose the initial analysis.

SUMMARY

In this chapter, we focused on the steps that come before training machine learning models. We discussed how to plan a machine learning strategy and learned about various hands-on methods we can use to prepare a dataset for modeling.

Starting with a high-level view, we focused on approaching data science problems by looking at available data, determining business needs, and assessing the data for suitability. Next, we discussed how to understand data from a modeling perspective, such as being able to identify whether datasets lend themselves to supervised or unsupervised learning problems.

Having covered these big-picture ideas, we paid particular attention to data preparation, which should be performed prior to modeling. We saw how to merge datasets, drop or fill missing values, transform categorical features, and split datasets into training and testing sets.

Finally, we introduced the Human Resource Analytics dataset and put what we learned into practice by cleaning it up for modeling. In the following chapters, we will use this processed dataset to train a variety of classification models. We'll start by introducing our modeling algorithms and overviewing how they work, and then use Jupyter to train and compare their predictive capabilities.

TRAINING CLASSIFICATION MODELS

OVERVIEW

In this chapter, you will learn about algorithms such as Support Vector Machines, Random Forests, and k-Nearest Neighbors classifiers. While training and comparing a variety of models, you'll learn about the concept of overfitting with the help of decision boundary charts. By the end of this chapter, you will be able to use scikit-learn to apply these algorithms in order to train models for a real-world classification problem.

INTRODUCTION

In the previous chapters, we walked through the steps that we need to take in a data science project before we can train a machine learning model. This included the planning phase, that is, identifying business problems, assessing data sources for suitability, and deciding on modeling approaches.

Having decided on a general modeling approach, we should be careful to avoid the **common pitfalls of training ML models** as we proceed with modeling. Firstly, remember that training data is very important. In fact, increasing the amount of training data can have a larger impact than model selection on scoring performance. One issue is that there may not be enough data available, which could make patterns difficult to find and cause models to perform poorly on testing data. **Data quality** also has a huge effect on model performance. Some possible issues include the following:

- Non-representative training data (sampling bias)

- Errors in the record sets (such as recorded weight in kg instead of pounds)

- Outliers

- Unfilled missing values

- Bad features

These issues should bring to mind the work that we did in the previous chapter, when preprocessing our datasets.

On the modeling side, you should avoid **overfitting**, which happens when the model fits too well on the training data so that it fails to generalize to testing samples. Similarly, you should also try to avoid **underfitting**, where the model is not able to capture the interesting patterns that would yield higher accuracy. The solution to this is trying many different types of models and being careful to follow best practices such as k-fold cross validation, as will be discussed in *Chapter 5, Model Validation and Optimization*. In order to prepare you for that, this chapter will focus on understanding what overfitting actually looks like, and the general concepts of how to avoid it.

Near the end of *Chapter 3, Preparing Data for Predictive Modeling*, we introduced the Human Resource Analytics dataset (https://www.kaggle.com/giripujar/hr-analytics). We'll continue to use this dataset in the exercises designed for this chapter, when training and comparing our models. Whereas previously, we loaded a messy version of the dataset and cleaned it up using pandas and Jupyter, in this chapter, we will load the cleaned-up version of the dataset.

This chapter is very hands-on, and the majority of what we'll learn will be from the exercises we'll perform when it comes to working with Jupyter. We'll do this by modeling employee turnover by training various classification algorithms using scikit-learn. First, however, we'll take a moment to briefly introduce how the algorithms work from a conceptual point of view.

UNDERSTANDING CLASSIFICATION ALGORITHMS

Recall the two types of supervised machine learning: regression and classification. In regression, we predict a numerical target variable. For example, recall the linear and polynomial models from *Chapter 2, Data Exploration with Jupyter*. Here, we will focus on the other type of supervised machine learning—classification— the goal of which is to predict the class of a record using the available metrics. In the simplest case, there are only two possible classes, which means we are doing binary classification. This is the case for the example problem in this chapter, where we will try to predict whether an employee is going to leave. If we have more than two class labels, then we are doing multi-class classification.

Although there is little difference between binary and multi-class classification when it comes to training models with scikit-learn, the algorithms can be notably different. In particular, multi-class classification models often use the one-versus-all method. This works as follows, for a case with three class labels. When the model is "fit" with the data, three models are trained, and each model predicts whether the record is part of an individual class or part of some other class. Then, when making a prediction, each model is evaluated and the class label with the highest confidence level is returned.

In this chapter, we'll train three types of classification models: **Support Vector Machines (SVMs)**, **Random Forests**, and **k-Nearest Neighbors classifiers (KNN)**. Each of these algorithms is quite different. As we will see, however, they are quite similar to train and use for predictions, thanks to the simplicity of scikit-learn. Before opening the Jupyter notebook for this chapter, let's briefly discuss how each of these algorithms works.

SVMs attempt to find the best hyperplane to divide classes by. This is done by maximizing the distance between the hyperplane and the closest records of each class, which are called support vectors.

While the basic implementation works best on linearly separable data, SVMs can also be used to model nonlinear dependencies by using the kernel trick. In short, the kernel trick maps dataset features into a higher-dimensional space. Once in this higher-dimensional space, the data is then assumed to be linearly separable so that a hyperplane can be found.

This hyperplane is also referred to as the decision surface, and we'll visualize it when training our models in the exercises in this chapter. In fact, we will see examples of both linear and non-linear (kernel) SVMs.

Random Forests are an ensemble of decision trees, where each has been trained on different subsets of the training data. For example, a Random Forest classifier might take the average result of many hundreds of decision trees in order to make a classification. A classification **decision tree** predicts the class of a given record based on a series of cascading decisions. For example, the first decision might be `"if feature x_1 is less than or greater than 0"`. The data would then be split on this condition and fed into descending branches of the tree. When "training" decision trees (that is, when feeding the training data into the algorithm), the details of each decision are selected based on the feature split that maximizes the information gain. Essentially, the algorithm attempts to separate the maximum number of class labels at each branch split so that each bucket at the bottom of the tree will contain training records with the same target label.

Training a Random Forest consists of creating bootstrapped datasets (that is, randomly sampled data with replacement so that a record can be duplicated multiple times in the training dataset) for a set of decision trees. Predictions are then made based on the majority vote. These have the benefit of less overfitting and better generalizability.

> **NOTE**
>
> Decision trees can be used to model a mix of continuous and categorical data, which make them very useful. Furthermore, as we will see later in this chapter, the tree depth can be limited to reduce overfitting. For a detailed (but brief) look into the decision tree algorithm, check out this popular StackOverflow answer, where the author shows a simple example and discusses concepts such as node purity, information gain, and entropy: https://stackoverflow.com/questions/1859554/what-is-entropy-and-information-gain/1859910#1859910.

KNN classification algorithms memorize the training data and make predictions depending on the k-nearest records in the feature space. With three features, this can be visualized as a sphere surrounding the prediction sample. Often, however, we will be dealing with more than three features, and therefore hyperspheres are drawn to find the closest k records.

Having introduced the algorithms we'll be using to train models in this book, we are ready to proceed with the hands-on section of this chapter.

EXERCISE 4.01: TRAINING TWO-FEATURE CLASSIFICATION MODELS WITH SCIKIT-LEARN

In this exercise, we'll continue working on the employee retention problem that we introduced in the previous chapter.

Recall the context of this problem, where a company has come to us asking for help with understanding and preventing employee turnover. To refresh your memory, you may wish to go back and read through the problem description and modeling plan from *Chapter 3, Preparing Data for Predictive Modeling*. As we describe there, our approach to helping with this problem was to determine the probability of an employee leaving the company. Using the Human Resource Analytics dataset provided, we planned out a strategy to model whether an employee is going to leave the company.

As seen in the dataset, this model can depend on an assortment of information (features) relating to the employee, such as satisfaction level, the number of projects they are working on, and their department in the company. To start with, however, we are going to focus on just two features: the satisfaction level and last evaluation score. This way, we can focus on the technical details of modeling and illustrate how different modeling approaches compare. In the next chapter, we will build upon the work we've done here by training models on the full set of features at our disposal. Perform the following steps to complete this exercise:

1. Create a new Jupyter notebook.

2. In the first cell, add the following lines of code to load the necessary libraries and set up your plot environment for the notebook:

```
import numpy as np
import pandas as pd
import datetime
import time
import os

import matplotlib.pyplot as plt
%matplotlib inline
import seaborn as sns
```

```
%config InlineBackend.figure_format='retina'
sns.set() # Revert to matplotlib defaults
plt.rcParams['figure.figsize'] = (8, 8)
plt.rcParams['axes.labelpad'] = 10
sns.set_style("darkgrid")
```

3. In the next cell, enter the following code to print the date, version numbers, and hardware information:

```
%load_ext watermark
%watermark -d -v -m -p \
requests,numpy,pandas,matplotlib,seaborn,sklearn
```

You should get the following output:

```
2020-02-13

CPython 3.7.5
IPython 7.10.1

requests 2.22.0
numpy 1.17.4
pandas 0.25.3
matplotlib 3.1.1
seaborn 0.9.0
sklearn 0.21.3

compiler   : Clang 4.0.1 (tags/RELEASE_401/final)
system     : Darwin
release    : 18.7.0
machine    : x86_64
processor  : i386
CPU cores  : 8
interpreter: 64bit
```

Figure 4.1: Screenshot showing all the libraries loaded

We are going to start by taking a closer look at our target variable, **left**, which has either of the following values:

1 (True) if the employee has left

0 (False) if the employee is still working at the company

In the previous chapter, you processed the raw dataset and saved your transformed table in a CSV file. You can look back to the bottom of the **chapter_3_workbook.ipynb** notebook to see when this was done.

4. Load the processed dataset by running the following code in a new cell:

```
df = pd.read_csv('[PATH_TO_THE_data_FOLDER]/hr-'\
                 'analytics/hr_data_processed.csv')
```

> **NOTE**
>
> It's easy to accidently introduce errors into your processed dataset while working through the data cleaning stage in *Chapter 3, Preparing Data for Predictive Modeling*. To avoid this possibility, or in case you are having issues working through the exercises in this chapter, make sure you are using the processed dataset provided in the source material for this book.
>
> The processed dataset is available to download from GitHub at https://packt.live/3hEBQly.

In this chapter, we'll be training classification models on two sets of continuous features: **satisfaction_level** and **last_evaluation**. Based on their histograms, which we charted out in *Chapter 3, Preparing Data for Predictive Modeling* (https://packt.live/2YEsiUX), we can see how they are distributed rather evenly between **0** and **1**. Now, let's look at how they are distributed with respect to one another.

5. Draw the bivariate (and univariate) graphs of the two feature variables by running the following code:

```
sns.jointplot(x='satisfaction_level', y='last_evaluation', \
              data=df, kind='hex')
plt.savefig('../figures/chapter-4-hr-analytics-jointplot.png', \
            bbox_inches='tight', dpi=300,)
```

You'll get the following output. As you can see, there are some very distinct patterns in the data:

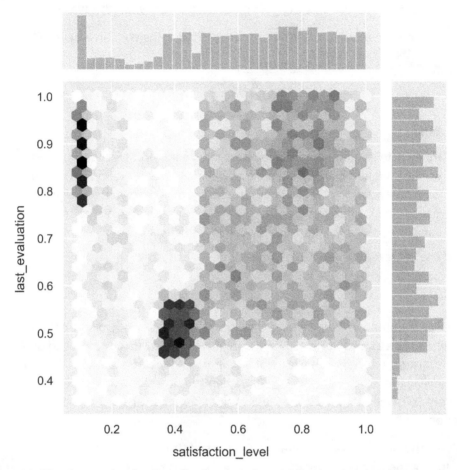

Figure 4.2: Bivariate and univariate distributions for satisfaction_level and last_evaluation

The preceding density chart is useful, but it would be even more interesting to see how this distribution differs when comparing employees who have left with those who have stayed in the company.

6. Replot the bivariate distribution, but this time, segment the chart on the target variable so that each target class is represented by a different color. This can be done with the following code:

```
fig, ax = plt.subplots()
plot_args = dict(shade=True, shade_lowest=False)
for i, c in zip((0, 1), ('Reds', 'Blues')):
```

```
        sns.kdeplot(df.loc[df.left==i, 'satisfaction_level'], \
                df.loc[df.left==i, 'last_evaluation'], \
                cmap=c, **plot_args)

ax.text(0.05, 1.05, 'left = 0', size=16, \
        color=sns.color_palette('Reds')[-2])
ax.text(0.25, 1.05, 'left = 1', \
        size=16, color=sns.color_palette('Blues')[-2])
plt.savefig('../figures/chapter-4-hr-analytics-bivariate-'\
            'segmented.png', bbox_inches='tight', dpi=300,)
```

You will see the segmentation of data. The output will be as follows:

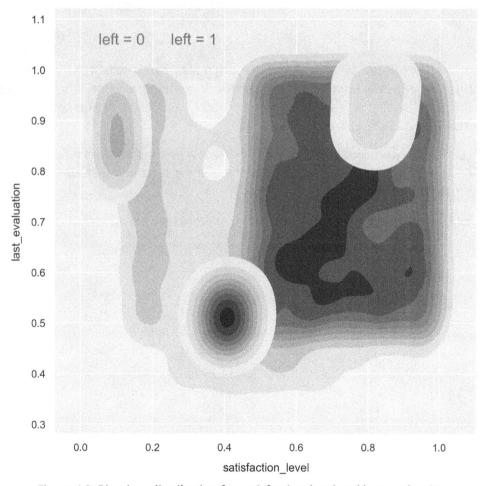

Figure 4.3: Bivariate distribution for satisfaction_level and last_evaluation, segmented by the target variable

Now, you can see how the patterns are related to the target variable. For the remainder of this exercise, try to exploit these patterns to train effective classification models.

7. Split the data into training and test sets by running the cell containing the following code:

```
from sklearn.model_selection import train_test_split


features = ['satisfaction_level', 'last_evaluation']
X_train, X_test, \
y_train, y_test = train_test_split(df[features].values, \
                                   df['left'].values, \
                                   test_size=0.3, \
                                   random_state=1)
```

As is often the case when it comes to machine learning, the first two models (the SVM and kNN classifiers) are most effective when the input data is scaled so that all of the features are in the same order. Accomplish this with scikit-learn's **StandardScaler** class

8. Load the **StandardScaler** class and create a new instance, as referenced by the **scaler** variable. Then, fit this on the training set and transform it, before finally transforming the test set:

```
from sklearn.preprocessing import StandardScaler


scaler = StandardScaler()
X_train_std = scaler.fit_transform(X_train)
X_test_std = scaler.transform(X_test)
```

> **NOTE**
>
> An easy mistake to make when training machine learning models is to *fit* the scaler on the whole dataset, when in fact it should only be *fit* to the training data. For example, scaling the data before splitting it into training and testing sets is a mistake.
>
> We don't want this to happen because then, the scaling of our training data will be affected by the details of our testing set, and we do not want the model training to be in any way influenced by the test data. By allowing the model training to be influenced by the test data, we would be giving it the opportunity to learn about the testing set, thereby causing it to be a poor representation of true "unseen" data.

9. Import the scikit-learn SVM class, SVC, and fit the model on the training data with the following code:

```
from sklearn.svm import SVC

svm = SVC(kernel='linear', C=1, random_state=1, gamma='scale')
svm.fit(X_train_std, y_train)
```

You will get the following output:

```
SVC(C=1, cache_size=200, class_weight=None, coef0=0.0,
    decision_function_shape='ovr', degree=3, gamma='scale',
    kernel='linear', max_iter=-1, probability=False,
    random_state=1, shrinking=True,
    tol=0.001, verbose=False)
```

10. Compute the accuracy of this model on unseen (test) data by running the following code:

```
from sklearn.metrics import accuracy_score

y_pred = svm.predict(X_test_std)
acc = accuracy_score(y_test, y_pred)
print('accuracy = {:.1f}%'.format(acc*100))
```

This will print the following output:

```
accuracy = 75.9%
```

Here, you have used scikit-learn's **accuracy_score** function, which takes in two arguments: the correct values of each record in the testing set and the predicted values of each record in the testing set (**y_test** and **y_pred**, respectively).

The result looks promising at ~75%, which is not bad for our first model. However, you may recall from *Chapter 3, Preparing Data for Predictive Modeling,* that our target class, **left**, is highly imbalanced, since the majority of the records in the dataset correspond to employees who are still working at the company. With this in mind, we need to think about the individual class accuracies, in addition to the overall accuracy.

11. To look at the accuracy within each target class, use a **confusion matrix**. This is a 2 x 2 table with actual classes on the horizontal axis and predicted classes on the vertical axis, as follows:

Predicted Class

	Class 0 (left = no)	Class 1 (left = yes)
Class 0 (left = no)	True Negative	False Positive
Class 1 (left = yes)	False Negative	True Positive

Actual Class

Figure 4.4: Confusion matrix for the linear SVM model

A perfect classifier would have all predictions as either True Positive or True Negative, and hence show zeros on the off-diagonal entries for False Positive and False Negative.

Print the confusion matrix for our model by running the following command:

```
from sklearn.metrics import confusion_matrix
cmat = confusion_matrix(y_test, y_pred)
```

This will print the following output:

```
array([[3416,    0],
       [1084,    0]])
```

Comparing this with the confusion reference table (*Figure 4.4*), you can see that our model is predicting all the test records as class 0, and hence we get over 1,084 False Negatives. This is not good!

12. Using the preceding confusion matrix, derive the class accuracies by running the following code:

```
print('percent accuracy score per class:')
cmat = confusion_matrix(y_test, y_pred)
scores = cmat.diagonal() / cmat.sum(axis=1) * 100
print('left = 0 : {:.2f}%'.format(scores[0]))
print('left = 1 : {:.2f}%'.format(scores[1]))
```

This will print the following output:

```
percent accuracy score per class:
left = 0 : 100.00%
left = 1 : 0.00%
```

As expected, you can see that your model is simply classifying every sample as 0, meaning that it's predicting that no employees in the test set will leave the company. Clearly, this is not helpful at all.

Let's use a contour plot to show the predicted class at each point in the feature space. This is commonly known as the decision regions plot.

13. Plot the decision regions using a helpful function from the **mlxtend** library, as follows:

```
from mlxtend.plotting import plot_decision_regions

N_samples = 200
X, y = X_train_std[:N_samples], y_train[:N_samples]
plot_decision_regions(X, y, clf=svm)
```

Here's what the plot for decision regions looks like in the output:

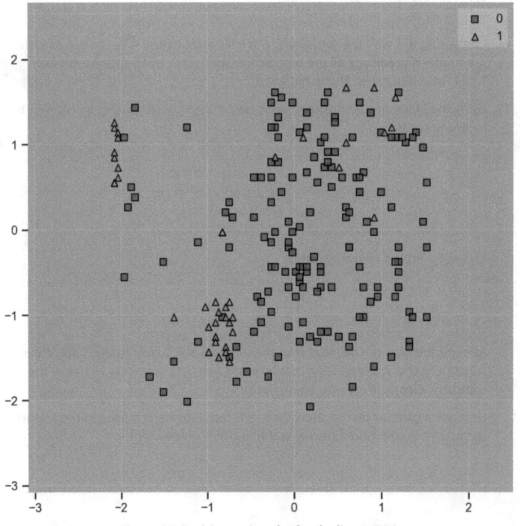

Figure 4.5: Decision region plot for the linear SVM

The preceding function plots decision regions, along with a set of records that are passed as arguments. In order to see the decision regions properly without too many records obstructing our view, we only pass a 200-record sample of the test data to the **plot_decision_regions** function. In this case, of course, it does not matter. We see the result is entirely red, indicating that every point in the feature space will be classified as 0.

It shouldn't be surprising that a linear model can't do a good job of describing these nonlinear patterns. If the dataset was linearly separable, we would be able to separate the majority of records from each class using a straight line. Recall the kernel trick for using SVMs to classify nonlinear problems. Let's see whether doing this can improve the result.

14. Print the docstring for scikit-learn's SVM by running the cell containing **SVC?**:

```
Init signature:
SVC(
    C=1.0,
    kernel='rbf',
    degree=3,
    gamma='auto_deprecated',
    coef0=0.0,
    shrinking=True,
    probability=False,
    tol=0.001,
    cache_size=200,
    class_weight=None,
    verbose=False,
    max_iter=-1,
    decision_function_shape='ovr',
    random_state=None,
)
Docstring:
C-Support Vector Classification.
```

Figure 4.6: The docstring for sklearn.svm.SVC

Scroll down and check out the parameter descriptions. This should read as follows:

```
kernel : string, optional (default='rbf')
    Specifies the kernel type to be used in the algorithm.
    It must be one of 'linear', 'poly', 'rbf', 'sigmoid',
    'precomputed' or a callable.
```

Notice the kernel option, which is enabled by default as **rbf**. This stands for the radial basis function, the details of which are beyond the scope of this book.

15. Use the **rbf** kernel option to train a new SVM by running the following code:

```
svm = SVC(kernel='rbf', C=1, random_state=1, gamma='scale')
svm.fit(X_train_std, y_train)
```

In order to continue rapidly building and comparing models, we are going to wrap the previous few steps into a Python function called **check_model_fit**, which will accept a trained model, along with the test data, and produce the accuracy scores and decision boundary chart we are interested in. Define this function by running the following code:

chapter_4_workbook.ipynb

```
from sklearn.metrics import accuracy_score
from sklearn.metrics import confusion_matrix
from IPython.display import display
from mlxtend.plotting import plot_decision_regions

def check_model_fit(clf, X_test, y_test):
    # Print overall test-set accuracy
    y_pred = clf.predict(X_test)
    acc = accuracy_score(y_test, y_pred, normalize=True) * 100
    print('total accuracy = {:.1f}%'.format(acc))
```

The complete code for this step can be found at https://packt.live/3e6JYPJ.

16. Call **check_model_fit** for your newly trained kernel SVM by running the following command:

```
check_model_fit(svm, X_test_std, y_test)
```

This will produce the following output:

```
total accuracy = 89.7%
            predictions

                 0     1

 actual  0   3308   108

         1    354   730
```

```
percent accuracy score per class:
left = 0 : 96.84%
left = 1 : 67.34%
```

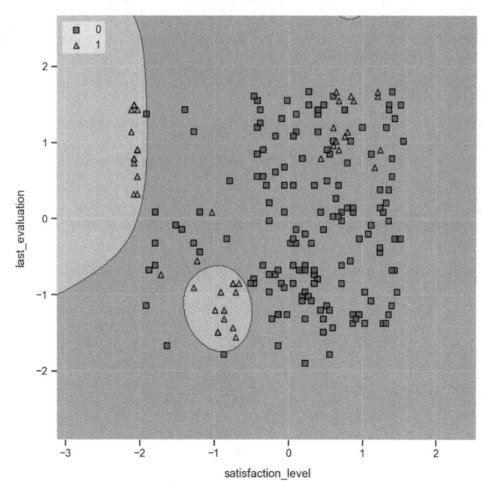

Figure 4.7: Decision region plot for the kernel SVM

This result is much better. We can see an overall accuracy of nearly 90%, where the class 1 accuracy is now 67%, compared to 0% with the linear SVM! We were able to capture the non-linear patterns in the data and correctly classify the majority of the employees who have left.

> ### NOTE
>
> To access the source code for this specific section, please refer to https://packt.live/30FSdOZ.
>
> You can also run this example online at https://packt.live/2ACdbUc.

In the remaining exercises for this chapter, we'll continue training models with scikit-learn and see how they compare to the SVMs from this section. Before doing this, however, we are going to briefly discuss how the decision boundaries are being plotted in the preceding charts.

THE PLOT_DECISION_REGIONS FUNCTION

In this section, we'll explore how Jupyter can help us look deeper into external Python library functions to learn how they work. In particular, we'll focus on the **plot_decision_regions** function that's provided by the **mlxtend** external library.

In the exercises for this chapter, we'll use the **plot_decision_regions** function to visualize how our models are learning the training data. It's worth taking a peek at the source code (which is written in Python) to understand how these plots are drawn:

```
from mlxtend.plotting import plot_decision_regions
plot_decision_regions?
```

By running the preceding code in a Jupyter notebook, the docstring can be seen as follows:

```
Signature:
plot_decision_regions(
    X,
    y,
    clf,
    feature_index=None,
    filler_feature_values=None,
    filler_feature_ranges=None,
    ax=None,
    X_highlight=None,
    res=None,
    zoom_factor=1.0,
    legend=1,
    hide_spines=True,
    markers='s^oxv<>',
    colors='#1f77b4,#ff7f0e,#3ca02c,#d62728,#9467bd,#8c564b,#e377c2,#7f7f7f,#bcbd22,#17becf',
    scatter_kwargs=None,
    contourf_kwargs=None,
    scatter_highlight_kwargs=None,
)
Docstring:
Plot decision regions of a classifier.

Please note that this functions assumes that class labels are
labeled consecutively, e.g,. 0, 1, 2, 3, 4, and 5. If you have class
labels with integer labels > 4, you may want to provide additional colors
and/or markers as `colors` and `markers` arguments.
See http://matplotlib.org/examples/color/named_colors.html for more
information.

Parameters
-----------
X : array-like, shape = [n_samples, n_features]
    Feature Matrix.
y : array-like, shape = [n_samples]
    True class labels.
clf : Classifier object.
    Must have a .predict method.
```

Figure 4.8: The docstring for mlxtend.plotting.plot_decision_regions

By scrolling to the bottom of the docstring printout in Jupyter, you can find the location of the file where this function is defined. This will look something like the following:

```
Returns
---------
ax : matplotlib.axes.Axes object

Examples
-----------
For usage examples, please see
http://rasbt.github.io/mlxtend/user_guide/plotting/plot_decision_regions/
File:          /anaconda3/lib/python3.7/site-packages/mlxtend/plotting/decision_regions.py
Type:          function
```

Figure 4.9: The file location and type, as seen at the bottom of the docstring in Jupyter

By taking note of the file path from the preceding screenshot, its contents can be printed in the notebook as follows (if bash is available):

```
%%bash
cat /anaconda3/lib/python3.7/site-packages/mlxtend/plotting/decision_
regions.py
```

The **bash** command gives us the following output:

```
# Sebastian Raschka 2014-2019
# mlxtend Machine Learning Library Extensions
#
# A function for plotting decision regions of classifiers.
# Author: Sebastian Raschka <sebastianraschka.com>
#
```

Figure 4.10: Printing the head of the plot_decision_regions source code

This allows us to view the code, but it's not very easy to look at since there's no color markup. A good solution is to copy it from the notebook (or copy the file itself) and open that in your favorite text editor. This can, in fact, be done directly inside Jupyter if you wish, since Jupyter functions as a plain text editor in addition to a notebook editor.

After this has been done, we can find the section of code that's responsible for creating the decision boundary chart. This is as follows:

```
214
215    xnum, ynum = plt.gcf().dpi * plt.gcf().get_size_inches()
216    xnum, ynum = floor(xnum), ceil(ynum)
217  1 xx, yy = np.meshgrid(np.linspace(x_min, x_max, num=xnum),
218                         np.linspace(y_min, y_max, num=ynum))
219
220    if dim == 1:
221        X_predict = np.array([[xx.ravel()]]).T
222    else:
223        X_grid = np.array([[xx.ravel(), yy.ravel()]]).T
224        X_predict = np.zeros((X_grid.shape[0], dim))
225        X_predict[:, x_index] = X_grid[:, 0]
226        X_predict[:, y_index] = X_grid[:, 1]
227        if dim > 2:
228            for feature_idx in filler_feature_values:
229                X_predict[:, feature_idx] = filler_feature_values[feature_idx]
230  2 Z = clf.predict(X_predict.astype(X.dtype))
231    Z = Z.reshape(xx.shape)
232    # Plot decisoin region
233    # Make sure contourf_kwargs has backwards compatible defaults
234    contourf_kwargs_default = {'alpha': 0.45, 'antialiased': True}
235    contourf_kwargs = format_kwarg_dictionaries(
236                          default_kwargs=contourf_kwargs_default,
237                          user_kwargs=contourf_kwargs,
238                          protected_keys=['colors', 'levels'])
239  3 cset = ax.contourf(xx, yy, Z,
240                       colors=colors,
241                       levels=np.arange(Z.max() + 2) - 0.5,
242                       **contourf_kwargs)
243
244    ax.contour(xx, yy, Z, cset.levels,
245               colors='k',
246               linewidths=0.5,
247               antialiased=True)
```

Figure 4.11: Interesting parts of the plot_decision_regions source code

By studying this code, we can see the following taking place, explained as per the annotated numbers:

- **1**: A mesh grid generated over the feature space.

- **2**: Predictions are then made at each point using scikit-learn's **predict** method on the classification model, **clf**.

- **3**: Finally, a contour plot is generated based on these predictions. This contour plot is what we see as output from the function when plotting the decision boundary charts.

It's not important to study and understand this code in detail. Instead, the purpose of this discussion was to pull back the curtains and show how to look deeper into the logic that Python libraries are performing.

In the next exercise, we'll return to modeling the employee retention problem that we started earlier.

EXERCISE 4.02: TRAINING K-NEAREST NEIGHBORS CLASSIFIERS WITH SCIKIT-LEARN

When training our first models on the Human Resource Analytics dataset, we saw how a linear SVM and kernel SVM performed dramatically differently when modeling two selected features from the dataset. Here, we will continue modeling these two features, **satisfaction_level** and **last_evaluation**, using the KNN algorithm. For the first time in this book, we'll visualize what overfitting looks like and learn about a strategy for handling it. Perform the following steps to complete this exercise:

1. Starting at the point in the notebook where the previous exercise ended, load the scikit-learn KNN classification model:

```
from sklearn.neighbors import KNeighborsClassifier
KNeighborsClassifier?
```

The preceding code will print the docstring:

```
Init signature:
KNeighborsClassifier(
    n_neighbors=5,
    weights='uniform',
    algorithm='auto',
    leaf_size=30,
    p=2,
    metric='minkowski',
    metric_params=None,
    n_jobs=None,
    **kwargs,
)
Docstring:
Classifier implementing the k-nearest neighbors vote.

Read more in the :ref:`User Guide <classification>`.

Parameters
----------
n_neighbors : int, optional (default = 5)
    Number of neighbors to use by default for :meth:`kneighbors` queries.
```

Figure 4.12: The docstring for sklearn.neighbors.KNeighborsClassifier

The **n_neighbors** parameter decides how many of the nearest-neighbor records to use when making a classification. If the **weights** argument is set to **uniform**, then class labels are decided by majority vote.

Another option you could consider for the **weights** argument is **distance**, where closer samples have a higher weight in the voting. Like most model parameters, the best choice for this depends on the particular dataset.

2. Train the KNN classifier with **n_neighbors=3**, and then compute the accuracy and decision regions. Since you have built all of this logic into the **check_model_fit** function, do this as follows:

```
knn = KNeighborsClassifier(n_neighbors=3)
knn.fit(X_train_std, y_train)
check_model_fit(knn, X_test_std, y_test)
```

Here's the output of the code:

```
total accuracy = 90.9%
            predictions

                0    1
  actual  0   3203  213
          1    198  886
```

```
percent accuracy score per class:
left = 0 : 93.76%
left = 1 : 81.73%
```

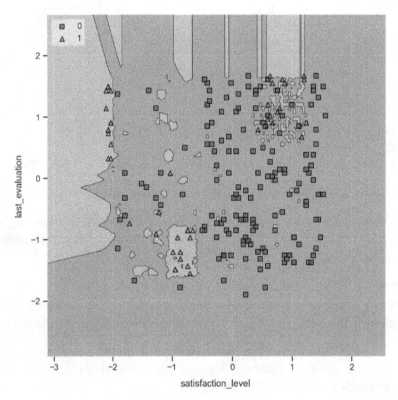

Figure 4.13: Training a KNN classifier with n_neighbors=3

Here, you can see an increase in overall accuracy, which now scores over 90%, and a significant improvement for class 1 in particular, which scores around 96%. However, the decision region plot indicates that we are overfitting the data. This is evident by the hard, *choppy* decision boundary, and small pockets of class 1 prediction ranges (the orange contours) scattered throughout the feature space. You can soften the decision boundary and decrease overfitting by increasing the number of nearest neighbors used to make classifications.

3. To reduce overfitting, train a KNN model with **n_neighbors=25** by running the following code:

```
knn = KNeighborsClassifier(n_neighbors=25)
knn.fit(X_train_std, y_train)
check_model_fit(knn, X_test_std, y_test)
```

You will see the following output:

```
total accuracy = 91.6%
            predictions
                 0     1

actual  0    3290   126
        1     254   830


percent accuracy score per class:
left = 0 : 96.31%
left = 1 : 76.57%
```

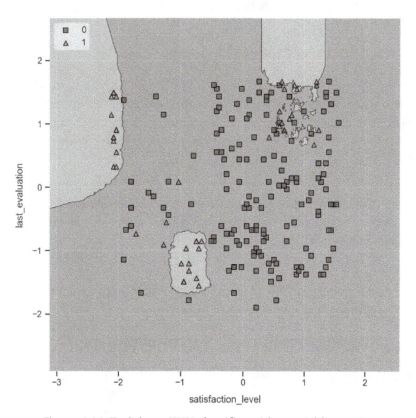

Figure 4.14: Training a KNN classifier with n_neighbors=25

As you can see, the decision boundaries are significantly less choppy compared to the plot for the **n_neighbors=3** model, and there are far fewer pockets of class 1 prediction ranges (the orange contours).

Looking at the metrics, the accuracy for class 1 is slightly less, but we would need to use a more comprehensive method (such as k-fold cross validation) to decide whether there's a significant difference between this model and the **n_neighbors=3** model that we trained previously.

Note that increasing **n_neighbors** has no effect on training time as the model is simply memorizing the data. The prediction time, however, becomes longer as **n_neighbors** is increased.

> **NOTE**
>
> To access the source code for this specific section, please refer to https://packt.live/30FSdOZ.
>
> You can also run this example online at https://packt.live/2ACdbUc.

> **NOTE**
>
> When using machine learning for real-world problems, it's important for the algorithms to run quick enough to serve their purposes. For example, a script to predict tomorrow's weather couldn't possibly require more than a day to run.
>
> Memory limits are also a consideration that should be taken into account when training models on substantial amounts of data. When this becomes an issue, you can look into models that can be iteratively trained on smaller chunks of the data.

Having finished training the KNN classifiers, we'll continue to explore modeling with scikit-learn by training Random Forests in the next exercise.

EXERCISE 4.03: TRAINING RANDOM FOREST CLASSIFIERS WITH SCIKIT-LEARN

We'll start this exercise right where the previous one left off. So far, we have trained SVMs and KNNs and seen what overfitting and underfitting looks like in terms of the decision boundary and resulting classification accuracy along the way.

Here, you will learn how to train Random Forests and compare their results for our modeling problem to those from the preceding algorithms. You will also learn how to render a decision tree visualization in the notebook, which will provide insights to the inner workings of our Random Forest models. Perform the following steps to complete this exercise:

1. Starting at the point in the notebook where the previous exercise ended load the scikit-learn Random Forest classification model and print the docstring by running the following command:

```
from sklearn.ensemble import RandomForestClassifier
RandomForestClassifier?
```

You will see the following output:

```
Init signature:
RandomForestClassifier(
    n_estimators='warn',
    criterion='gini',
    max_depth=None,
    min_samples_split=2,
    min_samples_leaf=1,
    min_weight_fraction_leaf=0.0,
    max_features='auto',
    max_leaf_nodes=None,
    min_impurity_decrease=0.0,
    min_impurity_split=None,
    bootstrap=True,
    oob_score=False,
    n_jobs=None,
    random_state=None,
    verbose=0,
    warm_start=False,
    class_weight=None,
)
Docstring:
A random forest classifier.

A random forest is a meta estimator that fits a number of decision tree
classifiers on various sub-samples of the dataset and uses averaging to
improve the predictive accuracy and control over-fitting.
The sub-sample size is always the same as the original
input sample size but the samples are drawn with replacement if
`bootstrap=True` (default).
```

Figure 4.15: The docstring for sklearn.ensemble.RandomForestClassifier

Here, scikit-learn considers **RandomForestClassifier** an ensemble algorithm, as can be seen from the preceding **import** statement, where it's loaded from the **ensemble** folder. Recall that earlier in this chapter we discussed the fact that Random Forests are actually a collection of other classifiers called decision trees. When training our Random Forest, we are actually training a set of decision trees. Similarly, when making predictions, we are actually feeding records into each decision tree of the Random Forest and computing the average result.

We will train a Random Forest classification model composed of 50 decision trees, each with a max depth of 5. Looking at the following code, notice how the Python commands are exactly analogous to the SVM and KNN models we trained earlier, despite each algorithm working so differently in scikit-learn. By setting the **max_depth=5** argument, as described by the docstring we printed out previously, we limit the maximum number of *consecutive* splits to 5. This will become clear when we look at a decision tree visualization later in this exercise.

2. Train the Random Forest classifier we just described by running the following code:

```
forest = RandomForestClassifier(n_estimators=50, \
                                max_depth=5, random_state=1,)
forest.fit(X_train, y_train)

check_model_fit(forest, X_test, y_test)
plt.xlim(-0.1, 1.2)
plt.ylim(0.2, 1.2)
plt.savefig('../figures/chapter-4-hr-analytics-forest.png', \
            bbox_inches='tight', \
            dpi=300,)
```

You will see a total accuracy of **92.0%**. Here's the output:

```
total accuracy = 92.0%
```

	predictions	
	0	**1**
actual **0**	3371	45
1	317	767

```
percent accuracy score per class:
left = 0 : 98.68%
left = 1 : 70.76%
```

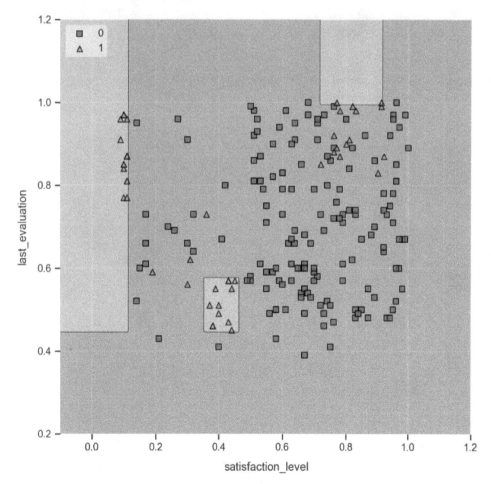

Figure 4.16: Training a Random Forest with a maximum depth of 5

While looking at the decision boundary chart, notice the distinctive axes-parallel decision boundaries produced by decision tree machine learning algorithms.

We can access any of the individual decision trees used to build the Random Forest. These trees are stored in the **estimators_** attribute of the model. Let's draw one of these decision trees to get a feel for what's going on.

> **NOTE**
>
> Producing decision tree graph visualizations with Python and scikit-learn requires the **graphviz** dependency, which can sometimes be difficult to install. If you are having difficulty with this step, you should open the **chapter-4-hr-analytics-tree-graph.png** graph file, which is available in the **figures** directory of the source code. This file can be found at the following link: https://packt.live/30IjVuh.

3. Build a graph for one of the decision trees from your Random Forest in the notebook by running the following code:

```
from sklearn.tree import export_graphviz
import graphviz

dot_data = export_graphviz(forest.estimators_[0], \
                           out_file=None, \
                           feature_names=features, \
                           class_names=['no', 'yes'], \
                           filled=True, rounded=True, \
                           special_characters=True,)
graph = graphviz.Source(dot_data)
```

4. Then, save the graph as a PNG file by running the following command:

```
graph.render(filename='../figures/chapter-4-hr-analytics-'\
             'tree-graph', format='png',)
```

5. Finally, render the graph in your notebook by running the **graph** command. This will return the following output:

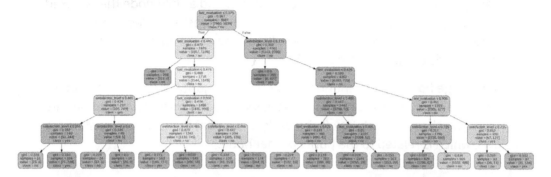

Figure 4.17: A decision tree from a Random Forest ensemble, where max_depth=5

From the preceding graph, we can see that each path is limited to five consecutive nodes as a result of setting **max_depth=5**. At each branch, scikit-learn's decision tree algorithm has decided on the feature split that maximizes the separability of classes in the training data. Consider the following section of the tree:

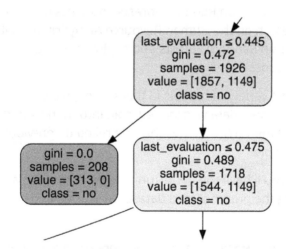

Figure 4.18: A section of the decision tree where a split is made
on the last_evaluation ≤ 0.445 condition

Here, we can see that 1,926 training samples from the top node have been split on the `last_evaluation` $\leq$ `0.445` condition, resulting in a child node that's pure (on the left) with 208 "no" samples, and a child node that's mixed (on the right) with 1,544 "no" samples and 1,149 "yes" samples. Recall that "no" corresponds to employees who are still working at the company, while "yes" corresponds to those who have left.

The orange boxes represent nodes where the majority of samples are labeled "no", and the blue boxes represent nodes where the majority of samples are "yes". The shade of each box (light, dark, and so on) indicates the confidence level, which is related to the purity of that node.

> **NOTE**
>
> To access the source code for this specific section, please refer to https://packt.live/30FSdOZ.
>
> You can also run this example online at https://packt.live/2ACdbUc.

This concludes our exercise on Random Forests and takes us to the end of our initial modeling research on the Human Resource Analytics dataset. In this exercise, we learned how to train Random Forests and explored how their decision tree constituents are composed.

Although we trained a variety of models in this section, we only worked through one end-to-end example where data was loaded, split into training and testing sets, used to train a model, and then scored. After that, we relied on previous work to make our modeling process simple.

In the next section, you'll have the opportunity to work through a full modeling activity, from loading the preprocessed dataset to scoring and comparing the final results.

ACTIVITY 4.01: TRAINING AND VISUALIZING SVM MODELS WITH SCIKIT-LEARN

In this activity, you'll build models in order to predict the same target variable that we described previously (whether or not an employee is going to leave) using two new features from the Human Resource Analytics dataset.

You'll select these features from the dataset, split them into training and testing sets, scale them, train and score some SVMs, and visualize the resulting decision boundaries.

It may be tempting to copy and paste code while completing this activity, but you will learn more effectively by typing out the solutions yourself. Even when referring back to earlier sections of the notebook, try to avoid copy and pasting. Perform the following steps to complete this activity:

> **NOTE**
>
> The detailed steps for this activity, along with the solutions and additional commentary, are presented on page 285.

1. Start a new Jupyter notebook, load the libraries you used, and set up your plotting environment for the notebook.

2. Load the preprocessed dataset from https://packt.live/2YE90iC (**hr_data_processed.csv**) into the notebook, assigning it to the **df** variable.

3. In this activity, you are going to use the **number_project** and **average_montly_hours** features. Filter **df** on these columns and display their summary metrics using the **pd.DataFrane.describe** function. Compare the mean, min, and max of each.

4. Do a train test split, as was done in *Exercise 4.01, Training Two-Feature Classification Models with Scikit-Learn*, using these two new features in place of the old ones.

5. Scale the training data using scikit-learn's **MinMaxScaler** and determine how this **scaler** differs from the one we used earlier in the notebook.

6. Train an SVM using the **rbf** kernel. Set the **C=1** and **gamma='scale'** arguments.

7. Identify the classification accuracy of this model for the test set.

8. Identify the class accuracies of this model for the test set.

9. Use the **plot_decision_regions** function to visualize the decision regions of this model.

10. When instantiating the SVM model, you passed in an argument, **C=1**. This is a property of the model that can be adjusted in order to optimize model accuracy. By increasing **C**, the SVM will attempt to fit the training data more closely. Train a new SVM with **C=50** and visualize its decision regions. Compare this to the chart you plotted for the **C=1** SVM.

SUMMARY

In this chapter, we learned about the SVM, KNN, and Random Forest classification algorithms and applied them to our preprocessed Human Resource Analytics dataset to build predictive models. These models were trained to predict whether an employee will leave the company, given a set of employee metrics.

For the purposes of keeping things simple and focusing on the algorithms, we built models that depend on only two features, that is, the satisfaction level and last evaluation value. This two-dimensional feature space also allowed us to visualize the decision boundaries and identify what overfitting looks like.

In the next chapter, we will introduce two important topics in machine learning: k-fold cross validation and validation curves. In doing so, we'll discuss more advanced topics, such as parameter tuning and model selection. Then, to optimize our final model for the employee retention problem, we'll explore feature extraction with the dimensionality reduction technique PCA, and train models on the full range of features available in the dataset.

5

MODEL VALIDATION AND OPTIMIZATION

OVERVIEW

In this chapter, you will learn how to use k-fold cross validation to test model performance, as well as how to use validation curves to optimize model parameters. You will also learn how to implement dimensionality reduction techniques such as **Principal Component Analysis** (**PCA**). By the end of this chapter, you will have completed an end-to-end machine learning project and produced a final model that can be used to make business decisions.

INTRODUCTION

As we've seen in the previous chapters, it's easy to train models with scikit-learn using just a few lines of Python code. This is possible by abstracting away the computational complexity of the algorithm, including details such as constructing cost functions and optimizing model parameters. In other words, we deal with a *black box* where the internal operations are hidden from us.

While the simplicity offered by this approach is quite nice on the surface, it does nothing to prevent the misuse of algorithms—for example, by selecting the wrong model for a dataset, overfitting on the training set, or failing to test properly on unseen data.

In this chapter, we'll show you how to avoid some of these pitfalls while training classification models and equip you with the tools to produce trustworthy results. We'll introduce k-fold cross validation and validation curves, and then look at ways to use them in Jupyter.

We'll also introduce the topic of dimensionality reduction and see how it can be used, along with k-fold cross validation, to perform model selection. We'll apply these techniques to our models for the Human Resource Analytics dataset in order to build and present an optimized final solution.

The topics in this chapter are highly practical with regard to real-world machine learning problems. The information and code presented here will enable you to build predictive models that perform well on unseen data, which is a crucial property of production models. To start things off, we'll learn about k-fold cross validation.

ASSESSING MODELS WITH K-FOLD CROSS VALIDATION

Thus far, we have trained models on a subset of the data and then assessed performance on the unseen portion, called the test set. This is good practice because the model's performance on data that's used for training is not a good indicator of its effectiveness as a predictor. It's very easy to increase accuracy on a training dataset by overfitting a model, which results in a poorer performance on unseen data.

That being said, simply training models on data that's been split in this way is not good enough. There is a natural variance in data that causes accuracies to be different (if even slightly), depending on the training and test splits. Furthermore, using only one training/test split to compare models can introduce bias toward certain models and lead to overfitting.

k-Fold cross validation offers a solution to this problem and allows the variance to be accounted for by way of an error estimate on each accuracy calculation.

The method of k-fold cross validation is illustrated in the following diagram, where we can see how the k-folds can be selected from the dataset:

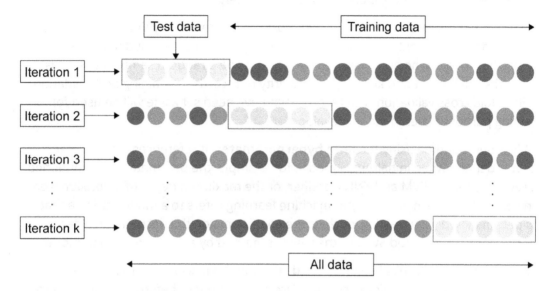

Figure 5.1: Illustration of k-fold cross validation

> **NOTE**
>
> Image source: CC BY-SA 4.0: https://commons.wikimedia.org/wiki/File:K-fold_cross_validation_EN.svg.

Keeping the preceding illustration in mind, the k-fold cross validation algorithm works as follows:

1. Split data into k *folds* of near-equal size.

2. Test and train k models on different fold combinations, where each model includes k – 1 folds of training data and uses the *left-out* fold as the validation set. In this method, each fold ends up being used as the test data exactly once.

3. Calculate the model accuracy by taking the mean of the k accuracy values. The standard deviation is also calculated to provide error estimates on the value.

It's standard to set *k = 10*, but smaller values for *k* should be considered if you're using a big dataset.

This validation method can be used to compare model performance with different hyperparameters in a more reliable way than using a single train-test split on the data, as we were doing in *Chapter 4, Training Classification Models*. For example, we could use k-fold cross validation to optimize the value of C for an SVM or the value of *k* (number of nearest neighbors) for a KNN classifier.

Although k-fold cross validation involves splitting the data many times into testing and validation sets, only a subset of the full dataset should be included in this algorithm. This can be accomplished by setting aside a random sample of records from the full dataset and keeping the majority or records for training and validation with k-fold cross validation. The records that have been set aside will be used for testing purposes later in model development.

This is our first time using the term **hyperparameter**. It references a parameter that is defined when initializing a model, for example, the parameters we mentioned previously for the SVM and KNN classifier, or the maximum depth of a decision tree. In contrast, the term *parameter* in machine learning refers to a model variable that is determined during training, such as the coefficients on the decision boundary hyperplane for a trained SVM, or the weights learned by a linear regression model.

Once the best model has been identified, it's often beneficial to retrain on the entirety of the dataset before using it in production (that is, before using it to make predictions). This way, we expose the model to all the data that's available so that it has an opportunity to learn from the full range of patterns in the dataset.

When implementing this with scikit-learn, it's common to use a slightly improved variation of the normal k-fold algorithm instead. This is called **stratified k-Fold**. The improvement is that stratified k-fold cross validation maintains roughly even class label populations in the folds. As you can imagine, this reduces the overall variance in the models and decreases the likelihood of highly unbalanced models causing bias.

TUNING HYPERPARAMETERS WITH VALIDATION CURVES

K-fold cross validation naturally lends itself to the use of validation curves for tuning model parameters. As shown in the following graph, validation curves chart out the model accuracy as a function of a hyperparameter, such as the number of decision trees used in a Random Forest or (as mentioned previously) the maximum depth. By understanding how to interpret these charts, we can make well-informed hyperparameter selections.

> **NOTE**
>
> Like most of the scikit-learn documentation, the information provided for validation curves is very informative and worth a read. This includes the recipes that we'll use in this chapter for creating and plotting validation curves. You can read about them at:
>
> https://scikit-learn.org/stable/modules/learning_curve.html.

Consider this validation curve, where the accuracy score is plotted as a function of the **gamma** SVM hyperparameter:

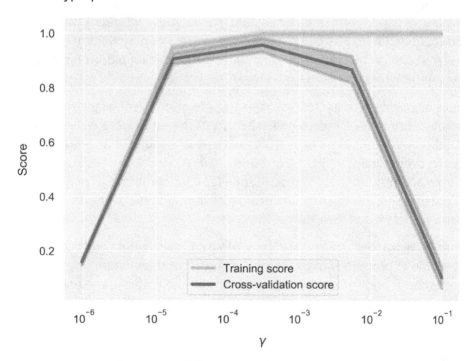

Figure 5.2: Validation curve for an SVM model

> **NOTE**
>
> Image source: https://scikit-learn.org/stable/auto_examples/model_selection/plot_validation_curve.html.

Starting on the left-hand side of the plot, we can see that both the training data (orange, top line) and the testing data (blue, bottom line) produce the same score. This is good, since it means the model is generalizing well to unseen data. However, the score is also quite low compared to other **gamma** values; therefore, we say the model is underfitting the data.

Increasing the value of **gamma**, we can see a point where the error bars of these two lines no longer overlap. From this point on, the classifier is failing to generalize well for unseen data, since it's overfitting the training set. As **gamma** continues to increase, we can see the score of the testing data drop off dramatically, while the training data score continues to increase.

The optimal value for the **gamma** parameter can be found by looking for a high test data score where the error bars on each line still overlap.

Keep in mind that a validation curve for some hyperparameters is only valid while the other hyperparameters remain constant. For example, if training the SVM in this plot, we could decide to pick gamma as 10^{-4}. However, we may want to optimize the C parameter as well. With a different value for C, the preceding plot would be different and our selection for gamma may no longer be optimal.

To handle problems such as this, you can look into **grid search algorithms**. These are available through scikit-learn and use many of the same ideas we've discussed here. Grid search works like a higher dimensional validation curve. For example, a grid search over `gamma = [10-5, 10-4, 10-3]` and `C = [0.1, 1, 10]` would train and test nine sets of models, one for each combination of gamma and C. As you can imagine, this is prone to becoming computationally intense as the number of hyperparameters and their ranges increase.

Now that we've learned the basics of how k-fold cross validation and validation curves work, it's time to proceed with the hands-on section of this chapter. We'll open up a Jupyter notebook and continue building models for the employee retention problem.

EXERCISE 5.01: USING K-FOLD CROSS VALIDATION AND VALIDATION CURVES IN PYTHON WITH SCIKIT-LEARN

In this exercise, you will implement k-fold cross validation and validation curves in Python and learn how to use these methods to assess models and tune hyperparameters in a Jupyter notebook. Perform the following steps to complete this exercise:

1. Create a new notebook using one of the following commands:

    JupyterLab (run **jupyter lab**)

    Jupyter Notebook (run **jupyter notebook**)

    Then, open the chosen platform in your web browser by copy and pasting the URL, as prompted in the Terminal.

2. Load the following libraries and set your plot setting for the notebook:

```
import pandas as pd
import numpy as np
import datetime
import time
import os

import matplotlib.pyplot as plt
%matplotlib inline
import seaborn as sns

%config InlineBackend.figure_format='retina'
sns.set() # Revert to matplotlib defaults
plt.rcParams['figure.figsize'] = (8, 8)
plt.rcParams['axes.labelpad'] = 10
sns.set_style("darkgrid")

%load_ext watermark
%watermark -d -v -m -p \
numpy,pandas,matplotlib,seaborn,sklearn
```

3. Start by loading the preprocessed training data (the same dataset we worked with in the previous chapter). Load the table by running the cell with the following code:

```
df = pd.read_csv('../data/hr-analytics/hr_data_processed.csv')
```

> **NOTE**
>
> As a reminder, you can find this file at https://packt.live/2YE90iC.

In this exercise, you will be working with the same two features as in the previous chapter: **satisfaction_level** and **last_evaluation**.

As mentioned previously in relation to k-fold cross validation, you still need to split the full dataset into a training and validation set and a test set. You will use the training and validation set during this exercise, and use the test set later during model selection.

4. Set up the training data by running the following command:

```
from sklearn.model_selection import train_test_split

features = ['satisfaction_level', 'last_evaluation']
X, X_test, \
y, y_test = train_test_split(df[features].values, \
                             df['left'].values, \
                             test_size=0.15, \
                             random_state=1)
```

5. Use a decision tree with **max_depth=5** to instantiate a model for k-fold cross validation

```
from sklearn.tree import DecisionTreeClassifier
clf = DecisionTreeClassifier(max_depth=5)
```

At this point, you have not performed any interesting computation. You have simply prepared a model object, **clf**, and defined its hyperparameters (for example, **max_depth**).

6. To run the stratified k-fold cross validation algorithm, use the **model_selection.cross_val_score** function from scikit-learn and print the resulting score by running the following code:

```
from sklearn.model_selection import cross_val_score

np.random.seed(1)
scores = cross_val_score(estimator=clf, X=X, \
                        y=y, cv=10,)

print('accuracy = {:.3f} +/- {:.3f}'.format(scores.mean(), \
                                            scores.std(),))
```

Here, you train **10** variations of our **clf** model using stratified k-fold validation. Note that scikit-learn's **cross_val_score** does this type of validation (stratified) by default.

You use **np.random.seed** to set the seed for the random number generator, thereby ensuring reproducibility with respect to any computation that follows that depends on random numbers. In this case, you set the seed to ensure reproducibility of the randomly selected samples for each fold in stratified k-fold cross validation.

Notice that you printed the average accuracy and standard deviation of each fold. You can also look at the individual accuracies for each fold by looking at the **scores** variable.

7. Insert a new cell and run **print(scores)**. You should see the following output:

```
[0.92241379 0.91529412 0.92784314 0.92941176 0.9254902  0.92705882
 0.91294118 0.91607843 0.92229199 0.9277865 ]
```

Using **cross_val_score** is a convenient way to accomplish k-fold cross validation, but it doesn't tell you about the accuracies within each class. Since your problem is sensitive to each class' accuracy (as identified in the exercises in the previous chapters), you will need to manually implement k-fold cross validation so that this information is available to us. In particular, you are interested in the accuracy of class 1, which represents the employees who have left.

8. Define a custom class for calculating k-fold cross validation class accuracies by running the following code:

```
from sklearn.model_selection import StratifiedKFold
from sklearn.metrics import confusion_matrix

def cross_val_class_score(clf, X, y, cv=10):
    kfold = (StratifiedKFold(n_splits=cv).split(X, y))
    class_accuracy = []
    for k, (train, test) in enumerate(kfold):
        clf.fit(X[train], y[train])
        y_test = y[test]
        y_pred = clf.predict(X[test])
        cmat = confusion_matrix(y_test, y_pred)
        class_acc = cmat.diagonal()/cmat.sum(axis=1)
        class_accuracy.append(class_acc)
        print('fold: {:d} accuracy: {:s}'.format(k+1, \
                                        str(class_acc),))
    return np.array(class_accuracy)
```

You implement k-fold cross validation manually using the **model_selection. StratifiedKFold** class in scikit-learn. This class takes the number of folds as an initialization argument and provides the split method to build randomly sampled masks for the data. In this instance, a **mask** is simply an array containing indexes of items in another array, where the items can then be returned by running code such as **data[mask]**.

9. Having defined this function, you can now calculate the class accuracies with code that's very similar to **model_selection.cross_val_score** from before. Do this by running the following code:

```
np.random.seed(1)
scores = cross_val_class_score(clf, X, y)

print('accuracy = {} +/- {}'.format(scores.mean(axis=0), \
                                scores.std(axis=0),))
```

This will print the following output when the folds are iterated over and 10 models are trained:

```
fold:  1 accuracy:  [0.98559671 0.72039474]
fold:  2 accuracy:  [0.98559671 0.68976898]
fold:  3 accuracy:  [0.98971193 0.72937294]
fold:  4 accuracy:  [0.98765432 0.74257426]
fold:  5 accuracy:  [0.99074074 0.71617162]
fold:  6 accuracy:  [0.98971193 0.72607261]
fold:  7 accuracy:  [0.98251029 0.68976898]
fold:  8 accuracy:  [0.98559671 0.69306931]
fold:  9 accuracy:  [0.98455201 0.72277228]
fold: 10 accuracy:  [0.98352214 0.74917492]
accuracy = [0.98651935 0.71791406] +/- [0.00266409 0.0200439 ]
```

These outputs show the class accuracies, where the first value corresponds to class **0** and the second corresponds to class **1**.

Having seen k-fold cross validation in action, we'll move on to the topic of validation curves. These can be generated easily with scikit-learn.

10. Calculate validation curves with **model_selection.validation_curve**. This function uses stratified k-fold cross validation to train models for various values of a specified hyperparameter. Perform the calculations required to plot a validation curve by running the following code:

```
from sklearn.model_selection import validation_curve

clf = DecisionTreeClassifier()
max_depth_range = np.arange(3, 20, 1)

np.random.seed(1)
train_scores, \
test_scores = validation_curve(estimator=clf, \
                               X=X, y=y, \
                               param_name='max_depth', \
                               param_range=max_depth_range, \
                               cv=5,);
```

By running this, you've trained a set of decision trees over the range of **max_depth** values. These values are defined in the **max_depth_range = np.arange(3, 20, 1)** line, which corresponds to the **[3, 4, ... 18, 19]** array—that is, from **max_depth=3** up to **max_depth=20**, with a step size of 1.

The **validation_curve** function will return arrays with the cross validation (training and test) scores for a set of models, where each has a different **max_depth** variable.

11. To visualize the results, leverage a function provided in the scikit-learn documentation:

> **NOTE**
>
> The triple-quotes (""") shown in the code snippet below are used to denote the start and end points of a multi-line code comment. Comments are added into code to help explain specific bits of logic.

chapter_5_workbook.ipynb

```
def plot_validation_curve(train_scores, \
                          test_scores, \
                          param_range, \
                          xlabel='', \
                          log=False, \
):
    """This code is from scikit-learn docs (BSD License).

    http://scikit-learn.org/stable/auto_examples/model_selection/plot_learning_
curve.html
    """
    train_mean = np.mean(train_scores, axis=1)
    train_std = np.std(train_scores, axis=1)
    test_mean = np.mean(test_scores, axis=1)
    test_std = np.std(test_scores, axis=1)
```

The complete code for this step can be found at https://packt.live/37vgad6.

This will result in the following graph:

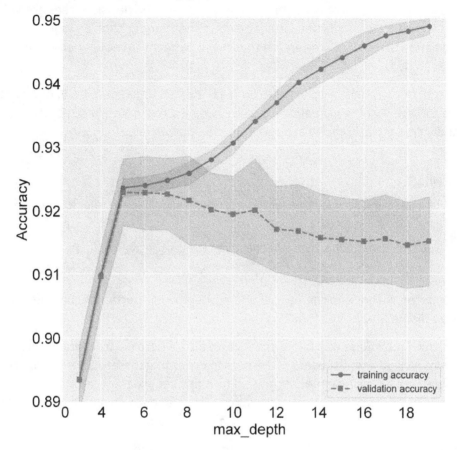

Figure 5.3: Validation curve for a decision tree

Setting the **max depth** parameter for decision trees controls the balance between underfitting and overfitting. This is reflected in the validation curve, where we can see low accuracies for small maximum depth values (underfitting), since we are not allowing the decision tree to create enough branches to capture the patterns in the data.

For large **max depth** values to the right of the chart, we can see the opposite happen, as the decision trees here overfit the training data. This is evidenced from the fact that our validation accuracy (red squares) decreases as the maximum depth increases.

Notice how the training accuracy (blue circles) continues increasing as the maximum depth increases. This happens because the decision trees are able to capture increasingly detailed patterns in the training data. By looking at the validation accuracies, we can see that these patterns do not generalize well for unseen data.

Based on this chart, a good value for **max_depth** appears to be **6**. At this point, we can see that the validation accuracy has hit a maximum and that the training and validation accuracies are agreement (within error).

> **NOTE**
>
> To access the source code for this specific section, please refer to https://packt.live/30GTi9a.
>
> You can also run this example online at https://packt.live/2BcG5tP.

To summarize, we have learned and implemented two important techniques for building reliable predictive models.

The first such technique was k-fold cross validation, where we train and validate a set of models over different subsets of the data in order to generate a variety of accuracy measurements for a single model choice. From this set, we then calculated the average accuracy and the standard deviation. This standard deviation is an important error metric to gauge the variability of our selected model.

The second technique we explored in this section was validation curves. By comparing training and validation accuracies (as generated by k-fold cross validation) over the range of our selected hyperparameter, validation curves allow us to visualize when our model is underfitting or overfitting and help us to identify optimal hyperparameter values.

In the next section, we'll introduce the concept of dimensionality reduction and why it's useful for training models. Then, we'll apply it to the Human Resource Analytics dataset and revisit the topics from this section in order to train highly accurate models for predicting employee turnover.

DIMENSIONALITY REDUCTION WITH PCA

Dimensionality reduction can be as simple as removing unimportant features from the training data. However, it's usually not obvious that removing a set of features will boost model performance. Even features that are highly noisy may offer some valuable information that models can learn from. For these reasons, we should know about better methods for reducing data dimensionality, such as the following:

- **Principal Component Analysis (PCA)**

- **Linear Discriminant Analysis (LDA)**

These techniques allow for data compression, where the most important information from a large group of features can be encoded in just a few features.

In this section, we'll focus on PCA. This technique transforms the data by projecting it into a new subspace of orthogonal principal components, where the components with the highest eigenvalues (as described here) encode the most information for training the model. Then, we can simply select a set of principal components in place of the original high-dimensional dataset. The number of principal components to select will depend on the details of the specific dataset, but it should be a reasonable percentage of the original set of features.

For example, PCA could be used to encode the information from every pixel in an image. In this case, the original feature space would have dimensions equal to the number of pixels in the image. This high-dimensional space could then be reduced with PCA, where the majority of useful information for training predictive models might be reduced to just a few dimensions. Not only does this save time when training and using models, but it allows them to perform better by removing noise from the dataset.

Similar to the algorithms we've discussed and implemented in this book, it's not necessary to have a detailed understanding of PCA in order to leverage its benefits. However, before implementing PCA with scikit-learn, we'll dig into the technical details a bit further in order to gain some appreciation for the underlying algorithm.

The key insight of PCA is to identify patterns between features based on correlations so that the PCA algorithm calculates the covariance matrix and then decomposes this into eigenvectors and eigenvalues. The vectors are then used to transform the data into a new subspace, from which a fixed number of principal components can be selected. Through this process, we effectively look at a high-dimensional dataset and find a set of vectors that follow directions of large variance, and thereby can encode much of the total information in fewer dimensions.

In the following exercise, we'll look at an example of how PCA can be used to reduce the dimensionality of our Human Resource Analytics dataset.

EXERCISE 5.02: DIMENSIONALITY REDUCTION WITH PCA

After training a variety of models for predicting employee turnover with the Human Resource Analytics dataset, we are still yet to use the majority of the features at our disposal. In this exercise, we will take the first steps in putting these features to use.

First, you will learn about a modeling technique that calculates which features are most influential for making predictions. Then, using these so-called "feature importance", you will create a strategy for selecting good features for dimensionality reduction. Finally, you will learn how to implement PCA with scikit-learn. Perform the following steps to complete this exercise:

1. Starting at the point in the notebook where the previous exercise ended, load the preprocessed dataset and print the columns by running the following code. This is the same table that you used in the previous exercise:

```
df = pd.read_csv('../data/hr-analytics/hr_data_processed.csv')
df.columns
```

Here's the output of the preceding command:

```
Index(['satisfaction_level', 'last_evaluation', 'number_project',
       'average_montly_hours', 'time_spend_company', 'work_accident', 'left',
       'promotion_last_5years', 'department_IT', 'department_RandD',
       'department_accounting', 'department_hr', 'department_management',
       'department_marketing', 'department_product_mng', 'department_sales',
       'department_support', 'department_technical', 'salary_high',
       'salary_low', 'salary_medium'],
      dtype='object')
```

Figure 5.4: The columns of hr_data_processed.csv

In order to determine which features are good candidates for reducing with PCA, you want to calculate how important each of them is for making predictions. Once you know this information, you can select those that are least important for PCA and leave the most important features intact.

2. Determine feature importance using a decision tree classifier. Select all available features and train a decision tree on the full dataset (not doing a train-test split), by running the following code:

```
features = ['satisfaction_level', 'last_evaluation', \
            'number_project','average_montly_hours', \
            'time_spend_company', 'work_accident', \
            'promotion_last_5years', 'department_IT', \
            'department_RandD','department_accounting', \
            'department_hr', 'department_management', \
            'department_marketing', 'department_product_mng', \
            'department_sales','department_support', \
            'department_technical', 'salary_high', \
            'salary_low', 'salary_medium']


X = df[features].values
y = df.left.values

from sklearn.tree import DecisionTreeClassifier
clf = DecisionTreeClassifier(max_depth=10)
clf.fit(X, y)
```

By now, you should recognize exactly what the preceding code is doing. Based on previous testing, you found **max_depth=6** to be a good choice when training on just two features: **satisfaction_level** and **last_evaluation**. When more features are included in the model, decision trees tend to require more depth to avoid underfitting (assuming all the other hyperparameters remain constant). Therefore, select **max_depth=10** as an educated guess. Most likely, this is not the optimal choice, but for our purposes here, this does not matter.

3. Having trained a *quick and dirty* model, leverage it to see how important each feature is for making predictions by using the **feature_importances_** attribute of **clf**. Visualize these in a bar chart by running the following code:

```
(
    pd.Series(clf.feature_importances_, \
             name='Feature importance', \
             index=df[features].columns,)
    .sort_values()
    .plot.barh()
)
plt.xlabel('Feature importance')
```

The bar plot for this is as follows:

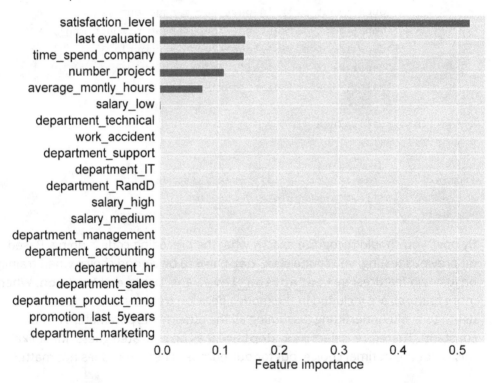

Figure 5.5: Feature importance calculated by a decision tree model

As shown in the preceding bar plot, there are a handful of features that are of significant importance when it comes to making predictions, and the rest appear to have near-zero importance.

Keep in mind, however, that this chart does not represent the *true* feature importance, but simply that of the *quick and dirty* decision tree model, `clf`. In other words, the features with near-zero importance in the preceding chart may be more important for other models. In any case, the information here is sufficient for us to make a selection on which features to reduce with PCA.

4. Set aside the five most important features from the preceding chart so that you can use them for modeling later, and then select the remainder for use in the PCA algorithm. Do this with the following code:

```
importances = list(pd.Series(clf.feature_importances_, \
                    index=df[features].columns,)
            .sort_values(ascending=False).index)
low_importance_features = importances[5:]
high_importance_features = importances[:5]
```

5. Print the least of low importance features list as follows:

```
np.array(low_importance_features)
```

The output is as follows:

```
array(['salary_low', 'department_technical', 'work_accident',
       'department_support', 'department_IT', 'department_RandD',
       'salary_high', 'salary_medium', 'department_management',
       'department_accounting', 'department_hr', 'department_sales',
       'department_product_mng', 'promotion_last_5years',
       'department_marketing'], dtype='<U22')
```

6. Print the least of high importance features list as follows:

```
np.array(high_importance_features)
```

The output is as follows:

```
array(['satisfaction_level', 'last_evaluation', 'time_spend_company',
       'number_project', 'average_montly_hours'], dtype='<U20')
```

7. Having identified the features to use for dimensionality reduction, run the PCA algorithm with the following code:

```
from sklearn.decomposition import PCA

pca_features = ['salary_low', 'department_technical', \
                'department_support','work_accident', \
                'salary_medium', 'department_IT', \
                'department_RandD', 'salary_high', \
                'department_management','department_accounting', \
                'department_hr', 'department_sales', \
                'department_product_mng', 'promotion_last_5years', \
                'department_marketing']

X_reduce = df[pca_features]

pca = PCA(n_components=3)
pca.fit(X_reduce)
X_pca = pca.transform(X_reduce)
```

First, we define the list of features to use in PCA, which can conveniently by done by copying and pasting the output of **np.array(low_importance_features)** from the preceding cell.

Next, we instantiate the PCA class from scikit-learn with **n_components=3**, indicating that we want to keep the first three components returned by the PCA algorithm. Finally, we fit our instantiated PCA class and then transform the same dataset.

8. Check the shape of the component data by running **X_pca.shape**. This will print the following output:

```
(14999, 3)
```

This result implies that we have three arrays of length 14,999 – corresponding to the three principal components for each record in the dataset.

9. Insert these principal component features into **df** by running the following code:

```
df['first_principle_component'] = X_pca.T[0]
df['second_principle_component'] = X_pca.T[1]
df['third_principle_component'] = X_pca.T[2]
```

10. Save the updated dataset by running the following code:

```
df.to_csv('../data/hr-analytics/hr_data_processed_pca.csv', \
          index=False,)
```

11. Finally, save the "fit" PCA class. This will be needed later to process future data before feeding it into our classifier's prediction method:

```
import joblib
joblib.dump(pca, 'hr-analytics-pca.pkl')
```

> **NOTE**
>
> To access the source code for this specific section, please refer to https://packt.live/30GTi9a.
>
> You can also run this example online at https://packt.live/2BcG5tP.

This concludes our exercise on PCA. You've learned how to generate feature importance and use that information to identify good candidates for dimensionality reduction. Using this technique, we found a set of features from the Human Resource Analytics dataset to apply the PCA algorithm on and reduced them to create three new features, representing their principal components.

MODEL TRAINING FOR PRODUCTION

So far in this book, we have trained many models and spent considerable effort learning about model assessment and optimization techniques. However, we have primarily focused on training models for instructional purposes, rather than producing production-ready models with optimal performance.

We have discussed the importance of training data various times in this book. Generally, we want to have as many training records and informative features as possible. One downside of having a massive set of records is the additional work required to clean that data in order to prepare it for use in machine learning algorithms. The same can be said for the number of features.

An additional problem that presents itself as the number of features grows is the difficulty in fitting models well. The variation of feature types, such as numerical, categorical, and Boolean, can restrict the type of models that are available to us and raise technical considerations around feature scaling during model training. In this chapter, we were able to avoid feature scaling altogether by using decision trees, which do not require features to be of a comparable scale.

More troubling than the preceding concerns, with respect to a growing number of features, is something known as the **curse of dimensionality**. This refers to the difficulty that models encounter when trying to fit a large number of features. As the number of dimensions in the training data increases, it becomes increasingly difficult for models to find patterns due to the inherently large distances that appear between records in a high-dimensional space. The dimensionality reduction techniques we learned about earlier can be effective for counteracting this effect.

Despite the difficulties outlined here, it still holds true that more training data is usually beneficial to model performance. So far in this book, we've worked mostly on training the majority of our models on just two features. In this section, we'll apply what we learned previously to model assessment and optimization in order to train a production-ready model that uses information from all of the features that are available in our dataset.

EXERCISE 5.03: TRAINING A PRODUCTION-READY MODEL FOR EMPLOYEE TURNOVER

We have already spent considerable effort planning a machine learning strategy, cleaning the raw data, and building predictive models for the employee retention problem. Recall that our business objective was to help the client prevent employees from leaving. The strategy we decided upon was to build a classification model that would be able to predict employee turnover by estimating the probability of an employee leaving. This way, the company can assess the likelihood of current employee turnover and take action to prevent it.

Given our strategy, we can summarize the type of predictive modeling we are doing as follows:

- Supervised learning on labeled training data

- Classification with two class labels (binary)

In particular, we are training models to determine whether an employee has left the company, given a set of numerical and categorical features.

After preparing the data for machine learning in *Chapter 3, Preparing Data for Predictive Modeling*, we went on to implement SVM, KNN, and Random Forest algorithms using just two features. These models were able to make predictions with over 90% overall accuracy. When looking at the specific class accuracies, however, we found that employees who had left (class label 1) could only be predicted with 70-80% accuracy.

In this exercise, you will see how much these class 1 accuracies can be improved by utilizing the full feature space. You will look at a unified example using validation curves for hyperparameter tuning, k-fold cross validation and test set verification for model assessment, as well as the final steps in preparing a production-ready model. Perform the following steps to complete this exercise:

1. Starting where you left off in the notebook, load the preprocessed dataset and print the columns by running the following code. This is the same table that you completed the previous exercise with:

```
df = pd.read_csv('../data/hr-analytics/hr_data_processed.csv')
df.columns
```

This command displays the following output:

```
Index(['satisfaction_level', 'last_evaluation', 'number_project',
       'average_montly_hours', 'time_spend_company', 'work_accident', 'left',
       'promotion_last_5years', 'department_IT', 'department_RandD',
       'department_accounting', 'department_hr', 'department_management',
       'department_marketing', 'department_product_mng', 'department_sales',
       'department_support', 'department_technical', 'salary_high',
       'salary_low', 'salary_medium'],
      dtype='object')
```

Figure 5.6: The columns of hr_data_processed.csv

As a quick refresher, we'll go through a brief summary of the variable descriptions. You are encouraged to look back at the analysis from *Chapter 3, Preparing Data for Predictive Modeling*, in order to review the feature distributions we generated.

The first two features, **satisfaction_level** and **last_evaluation**, are numerical and span continuously from 0 to 1; these are what we used to train the models in the previous two exercises. Next, we have some numerical features, such as **number_project** and **time_spend_company**, followed by Boolean fields such as **work_accident** and **promotion_last_5years**. We also have the one-hot encoded categorical features, such as **department_IT** and **salary_medium**. Lastly, we have the PCA variables representing the first three principal components of the select feature set from the previous exercise.

Given the mixed data types of our feature set, decision trees or Random Forests are very attractive models since they work well with feature sets composed of both numerical and categorical data. In this exercise, we are going to train a decision tree model.

> **NOTE**
>
> If you're interested in training an SVM or KNN classifier on mixed-type input features, you may find the data scaling prescription from this StackExchange answer useful: https://stats.stackexchange.com/questions/82923/mixing-continuous-and-binary-data-with-linear-svm/83086#83086.
>
> A simple approach would be to preprocess the data as follows:
>
> Standardize continuous variables, one-hot encode categorical features, and then shift binary values to -1 and 1 instead of 0 and 1.
>
> This would yield the data of mixed-feature types, which could then be used to train a variety of classification models.

2. Select the features to use for our model as the top five features from the PCA section, in terms of feature importance, and the first three principal components of the remaining features. Do this selection and split the data into a training and validation set (**X, y**) and a test set (**X_test, y_test**) by running the following code:

```
from sklearn.model_selection import train_test_split


features = ['satisfaction_level', 'last_evaluation', \
            'time_spend_company','number_project', \
            'average_montly_hours', 'first_principle_component', \
            'second_principle_component', \
            'third_principle_component',]
X, X_test, \
y, y_test = train_test_split(df[features].values, \
                             df['left'].values, \
                             test_size=0.15, \
                             random_state=1)
```

Notice that you set **test_size=0.15** for the train-test split, since you want to set aside 15% of the full dataset for testing the model that you will select after hyperparameter tuning.

The hyperparameter you are going to optimize is the decision tree's **max_depth**. You will do this in the same way you found the validation curves, where you found **max_depth=6** to be optimal for a decision tree model with only two features.

3. Calculate the validation curve for a decision tree with a maximum depth ranging from 2 up to 52 by running the following code:

```
%%time
from sklearn.tree import DecisionTreeClassifier


np.random.seed(1)
clf = DecisionTreeClassifier()
max_depth_range = np.arange(2, 52, 2)
print('Training {} models ...'.format(len(max_depth_range)))
train_scores, \
test_scores = validation_curve(estimator=clf, X=X, y=y, \
                               param_name='max_depth', \
                               param_range=max_depth_range, \
                               cv=10,);
```

Since you are using the **%%time** magic function, this cell will print a message similar to the following:

```
Training 25 models ...
CPU times: user 7.93 s, sys: 29.5 ms, total: 7.96 s
Wall time: 7.98 s
```

The details of this will depend on your hardware and the other processes happening on your system at runtime.

By executing this code, you run 25 sets of k-fold cross validation—one for each value of the **max_depth** hyperparameter in our defined range. By setting **cv=10**, you produce 10 estimates of the accuracy for each model (during k-fold cross validation), from which the mean and standard deviation are calculated in order to plot in the validation curve. In total, you train 250 models over various maximum depths and subsets of the data.

4. Having run the calculations required for the validation curve, plot it with the **plot_validation_curve** function that was defined earlier in the notebook. If needed, scroll up and rerun that cell to define the function. Then, run the following code:

```
plot_validation_curve(train_scores, test_scores, \
                    max_depth_range, xlabel='max_depth',)
plt.ylim(0.95, 1.0)
```

This will result in the following curve:

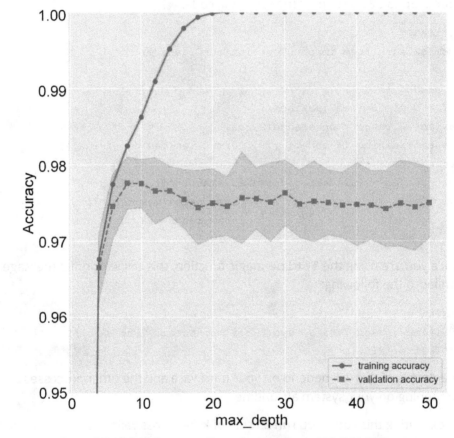

Figure 5.7: Validation curve for a decision tree with PCA features

Looking at this validation curve, you can see the accuracy of the training set (blue circles) quickly approach 100%, hitting this mark around **max_depth=20**. The validation set (red squares) reaches a maximum accuracy around **max_depth=8**, before dropping slightly as **max_depth** increases beyond this point. This happens because the models in this range are overfitting on the training data, learning patterns that don't generalize well to unseen data in the validation sets.

Based on this result, we can select **max_depth=8** as the optimal value to use for our production model.

5. Check the k-fold cross validation accuracy for each class in our model with the **cross_val_class_score** function that was defined earlier in the notebook. If needed, scroll up and rerun that cell to define the function. Then, run the following code:

```
clf = DecisionTreeClassifier(max_depth=8)
np.random.seed(1)
scores = cross_val_class_score(clf, X, y)

print('accuracy = {} +/- {}'.format(scores.mean(axis=0), \
                                    scores.std(axis=0),))
```

This will print the following output:

```
fold: 1 accuracy: [0.99382716 0.91118421]
fold: 2 accuracy: [0.99588477 0.91089109]
fold: 3 accuracy: [0.99897119 0.91749175]
fold: 4 accuracy: [0.99588477 0.95379538]
fold: 5 accuracy: [0.99279835 0.91419142]
fold: 6 accuracy: [0.99588477 0.92079208]
fold: 7 accuracy: [0.99485597 0.92409241]
fold: 8 accuracy: [0.99382716 0.9339934 ]
fold: 9 accuracy: [0.9907312  0.91419142]
fold: 10 accuracy: [0.99176107 0.94059406]
accuracy = [0.99444264 0.92412172] +/- [0.00226594 0.01357943]
```

As can be seen, this model is performing much better than previous models for class 1, with an average accuracy of 92.4% +/- 1.4%. This can be attributed to the additional features we are using here, compared to earlier models that relied on only two features.

6. Visualize the new class accuracies with a boxplot by running the following code:

```
fig = plt.figure(figsize=(5, 7))
sns.boxplot(data=pd.DataFrame(scores, columns=[0, 1]), \
                          palette=sns.color_palette('Set1'),)
plt.xlabel('Left (0="no", 1="yes")')
plt.ylabel('Accuracy')
```

Here's the visualization created by the preceding code:

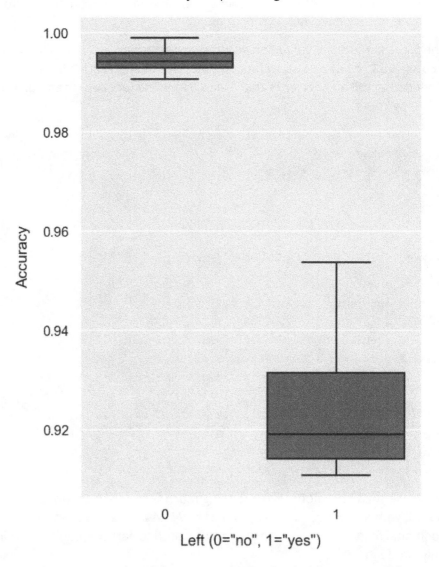

Figure 5.8: Boxplot of class accuracies for the decision tree model

At this point, having finished the hyperparameter optimization, you should now check how well the model performs on the test set. This result will give you more confidence that it will perform well when making predictions in production.

7. Train a model on the full set of training and validation data (**X, y**). Then, determine the accuracy of each class for the test set (**X_test, y_test**) by running the following code:

```
from sklearn.metrics import confusion_matrix

clf = DecisionTreeClassifier(max_depth=8)
clf.fit(X, y)

y_pred = clf.predict(X_test)
cmat = confusion_matrix(y_test, y_pred)
cmat.diagonal() / cmat.sum(axis=1) * 100
```

This will print the following output:

```
array([99.23976608, 93.88888889])
```

These test accuracies should fall within or very close to the range of the k-fold cross validation accuracies we calculated previously. For class 0, you can see 99.2%, which falls within the k-fold range of 99.2% – 99.6%, and for class 1, you can see 93.9%, which falls just above the k-fold range of 91.0% – 93.8%. These are good results, which give you confidence that your model will perform well in production.

8. You have nearly finished creating your production model. Having selected the best hyperparameters and tested the accuracy, now train a new model on the full dataset with the following code:

```
features = ['satisfaction_level', 'last_evaluation', \
            'time_spend_company', 'number_project', \
            'average_montly_hours', 'first_principle_component', \
            'second_principle_component', \
            'third_principle_component',]
X = df[features].values
y = df['left'].values

clf = DecisionTreeClassifier(max_depth=8)
clf.fit(X, y)
```

9. To use this model in production without needing to retrain it each time, save it to disk. Using the **joblib** module, dump the model to a binary file by running the following code:

```
import joblib
joblib.dump(clf, 'hr-analytics-pca-tree.pkl')
```

10. Check that your trained model was saved into the working directory. If your Jupyter Notebook environment has bash support, this can be done by running the following code:

```
!ls .
```

This will print the contents of the working directory:

```
chapter_5_workbook.ipynb     hr-analytics-pca-tree.pkl
hr-analytics-pca.pkl
```

11. In order to use this model to make predictions, load it from this binary file by running the following code:

```
clf = joblib.load('hr-analytics-pca-tree.pkl')
clf
```

The output of this command is as follows:

```
DecisionTreeClassifier(class_weight=None, criterion='gini', max_depth=8,
                max_features=None, max_leaf_nodes=None,
                min_impurity_decrease=0.0, min_impurity_split=None,
                min_samples_leaf=1, min_samples_split=2,
                min_weight_fraction_leaf=0.0, presort=False,
                random_state=None, splitter='best')
```

Figure 5.9: The decision tree model's representation

Now run through an example showing how this model can be used to make predictions regarding employee turnover. You will pick a record from the training data and feed it into the model for prediction.

12. Select a record from the training data and filter it on the original feature columns, and pretend this is the employee profile for Bob. Do this by running the following code:

```
pca_features = ['salary_low', 'department_technical', \
                'work_accident','department_support', \
                'department_IT', 'department_RandD', \
                'salary_high', 'salary_medium', \
                'department_management','department_accounting', \
                'department_hr', 'department_sales', \
                'department_product_mng', 'promotion_last_5years', \
                'department_marketing']

non_pca_features = ['satisfaction_level', 'last_evaluation', \
                    'time_spend_company','number_project', \
                    'average_montly_hours']

bob = df.iloc[8483][pca_features + non_pca_features]
bob
```

This will print the following output, showing Bobs metrics across each employee metric:

```
salary_low                 1.00
department_technical       0.00
work_accident              0.00
department_support         0.00
department_IT              0.00
department_RandD           0.00
salary_high                0.00
salary_medium              0.00
department_management      0.00
department_accounting      0.00
department_hr              0.00
department_sales           1.00
department_product_mng     0.00
promotion_last_5years      0.00
department_marketing       0.00
```

```
satisfaction_level           0.77
last_evaluation              0.68
time_spend_company           2.00
number_project               3.00
average_montly_hours       225.00
Name: 8483, dtype: float64
```

In general, a prediction sample would need to be prepared in exactly the same way that the training data was, which includes the same method of data cleaning such as filling missing values and one-hot encoding categorical variables.

In this case (for Bob), this preprocessing has already been done. However, at this point, assume that PCA transformations have not been done yet. This is a necessary step in order to produce the proper input that your model requires.

13. Load the PCA transformation class that was saved to disk earlier in this exercise and use it to transform the relevant features for Bob by running the following code:

```
pca = joblib.load('hr-analytics-pca.pkl')
pca_feature_values = pca.transform([bob[pca_features]])[0]
pca_feature_values
```

This will print the following output, showing the principal components that we need in order to make a prediction for Bob:

```
array([-0.67733089,  0.75837169, -0.10493685])
```

14. Create a prediction vector for Bob that can be input into the prediction method of your classification model by running the following code:

```
X_bob = np.concatenate((bob[non_pca_features].values, \
                        pca_feature_values))
X_bob
```

This will print the following output:

```
array([ 7.70000000e-01,  6.80000000e-01,  2.00000000e+00,
        3.00000000e+00,  2.25000000e+02, -6.77330887e-01,
        7.58371688e-01, -1.04936853e-01])
```

15. You are finally ready to see whether the model is predicting that Bob will leave the company. Calculate this outcome by running the following code:

```
clf.predict([X_bob])
```

This will print the following output:

```
array([0])
```

This indicates that our model is predicting that Bob will not leave the company, since he was assigned to class 0.

16. You can see what probability the model has assigned to this prediction by using its **predict_proba** method. Check this result by running the following code:

```
clf.predict_proba([X_bob])
```

This will print the following output:

```
array([[0.98, 0.02]])
```

This indicates that our model has assigned 98% probability to Bob remaining at the company.

> **NOTE**
>
> To access the source code for this specific section, please refer to https://packt.live/30GTi9a.
>
> You can also run this example online at https://packt.live/2BcG5tP.

You have now reached the end of our final exercise with the Human Resource Analytics dataset and successfully trained a model that can predict employee turnover. In this exercise, you used validation curves for hyperparameter optimization and k-fold cross validation model assessment to confirm the confidence in the model.

By training a model on the most important features, in addition to those produced from dimensionality reduction, we were able to build a model that performs much better than previous ones from *Chapter 3, Preparing Data for Predictive Modeling*.

Finally, you learned how to persist models on disk and reload them for use in making predictions.

In the following activity, you will attempt to improve on the model we trained here. This will give you an opportunity to apply the topics from this chapter and use the skills you have learned from this book.

ACTIVITY 5.01: HYPERPARAMETER TUNING AND MODEL SELECTION

In this final activity related to machine learning, we'll take everything we have learned so far and put it together in order to build another predictive model for the employee retention problem. We seek to improve the accuracy of the model from the preceding exercise by training a Random Forest model.

In order to accomplish this, you will need to use the methods you've seen being implemented throughout this chapter, such as k-fold cross validation and validation curves. You will also need to confirm the validity of your model on testing data and determine whether it's an improvement on previous work. Finally, you will apply the model to a practical business situation. Perform the following steps to complete this activity:

> **NOTE**
>
> The detailed steps for this activity, along with the solutions and additional commentary, are presented on page 293.

1. Start up one of the following platforms for running Jupyter Notebooks:

   JupyterLab (run **jupyter lab**)

   Jupyter Notebook (run **jupyter notebook**)

   Then, open the platform you chose in your web browser by copying and pasting the URL, as prompted in the Terminal.

2. Load the required libraries and set up your plotting environment for the notebook.

3. Start by loading the training dataset you generated earlier in the notebook (**hr_data_processed_pca.csv**), assigning it to the **df** variable.

4. Select the same features from the table that were used in the final exercise of this chapter (when we trained a decision tree with **max_depth=8**). Then, split these into a training and validation set and a test set (**X, X_test, y, y_test**). The test set should include 15% of the records.

5. Calculate a validation curve for Random Forest classification models with **n_estimators=50** over a range of **max_depth** values from 2 up to 52, in increments of 2. In the k-fold cross validation step of the validation curve calculation, assign the value of **k** to 5 by setting **cv=5**.

6. Draw the validation curve using the **plot_validation_curve** visualization that was defined earlier in the notebook. Interpret the chart and note anything that's different from the validation curve in the previous exercise. What would you pick as the optimal value for **max_depth**?

7. Perform k-fold cross validation using the **cross_val_class_score** function you defined earlier in the notebook, setting the Random Forest hyperparameters as **n_estimators=50** and **max_depth=25**. Are these results better than the decision tree we trained in the previous exercise?

8. Evaluate the performance of this model on the test set by training it on the full test and validation set (**X, y**), and then calculating its accuracy on each class in the test set (**X_test, y_test**). Are the scores in an appropriate range to validate the model?

9. Train this model on the full set of records in **df**.

10. Save the model to disk, and then check that it is saved properly by reloading it.

11. Check the model performance for an imaginary employee, Alice, by selecting the appropriate features from row 573 of **df**. Make sure you select all of the features needed to make a prediction, including the first, second, and third principal components.

12. Predict whether Alice is going to leave the company. Then, determine the probability assigned to that prediction by the model.

13. Adjust the feature values for Alice in order to determine the changes that would be required to alter the model's prediction. Try setting **average_montly_ hours=100** and **time_spend_company=2**. Then, rerun the model's prediction probabilities. Was this adjustment enough to sway the model's prediction on whether or not Alice is going to leave?

SUMMARY

In this chapter, we have seen how to use Jupyter Notebooks to perform parameter optimization and model selection.

We built upon the work we did in the previous chapter, where we trained predictive classification models for our binary problem and saw how decision boundaries are drawn for SVM, KNN, and Random Forest models. We improved on these simple models by using validation curves to optimize parameters and explored how dimensionality reduction can improve model performance as well.

Finally, at the end of the last exercise, we explored how the final model can be used in practice to make data-driven decisions. This demonstration connects our results back to the original business problem that inspired our modeling problem initially.

In the next chapter, we will depart from machine learning and focus on data acquisition instead. Specifically, we will discuss methods for extracting web data and learn about HTTP requests, web scraping with Python, and more data processing with pandas. These topics can be highly relevant to data scientists, given the huge importance of having good quality data to study and model.

6

WEB SCRAPING WITH
JUPYTER NOTEBOOKS

OVERVIEW

In this chapter, you will learn to make HTTP requests and parse data
from HTML. Like in previous chapters, you will continue to get hands-on
experience working with datasets in Python, including merging tables and
preparing them for analysis. By the end of this chapter, you will be able to
use Python to make HTTP requests, such as API calls, and create pipelines
to extract data from web pages.

INTRODUCTION

So far in this book, we have focused on using Jupyter to build reproducible data analysis and modeling workflows. We'll continue with a similar approach in this chapter, but with the main focus being on data acquisition. In particular, we will show you how data can be acquired from the web using HTTP requests. This will involve making API requests and scraping web pages by parsing HTML. In addition to these new topics, we'll continue to use pandas for building and transforming our datasets.

Before we cover HTTP requests and how to use them in Python, we'll discuss the importance of gathering data from the web in general. The amount of data that's available online is huge, and it's continuously growing at a staggering pace. Additionally, it's becoming increasingly important for driving business growth. Consider, for example, the ongoing global shift from technologies such as newspapers, magazines, and TV to online content. With customized newsfeeds available all the time on cell phones, and live news sources such as Facebook, Reddit, Twitter, and YouTube, it's easy to see why the historical alternatives continue to lose market share.

INTERNET DATA SOURCES

As data scientists, the internet helps connect us with any kind of dataset we could imagine. For instance, governments around the world publish public datasets that are rich with information. Along the same lines, some companies make certain datasets public, which can be of huge value within a given industry. One example of this is the ride-sharing business *Lyft*, who has released open source data that could be beneficial for training autonomous vehicles.

In addition to online datasets, **Application Programming Interface (API)** services also exist, which provide relevant and fresh data programmatically. For example, a business that depends on the weather may want an API that provides the current conditions in a given region, along with updated forecasts. Processes could be set up to query that API daily and update an internal database that's connected to a dashboard in order to provide that and other relevant data to business stakeholders.

Web scraping is the process of extracting information from a web page, or set of pages, using computer programs. This can be useful for pulling information from websites that have not made their data easily accessible through other means (that is, structured datasets or APIs).

While APIs are intended to be used programmatically, web scraping involves using computers to ingest data that was intended for humans to see and understand. Whereas API data is delivered in machine-readable formats such as XML or JSON, web pages will generally render data in a human-readable format using HTML.

Thus, web scraping often involves the ability to extract structured data from HTML text elements.

The exact process of web scraping will depend on the tool you're using, the specific web page, and the desired content. In this book, we'll learn about web scraping techniques we can use with Python by requesting and parsing data from Wikipedia. By focusing on the underlying concepts as we progress through, we'll learn how easy it can be to extract whatever is needed from an HTML page.

A more difficult aspect of web scraping can be requesting the HTML itself. It's understandable why companies would not want arbitrary programs to be able to request pages from their website since sending web pages to users requires their resources. Unlike humans who browse websites at a reasonable pace, computers are capable of asking for many web pages in a very small amount of time. Furthermore, programmatic traffic is not of any value to a website. For example, computers are not influenced by ads or interested in making a purchase.

> **NOTE**
>
> There may be legal issues with web scraping, depending on your region and the terms and conditions of the website. Please do research the rules and regulations with respect to your local jurisdiction before requesting web pages programmatically (that is, before doing the exercises and activity in this chapter). To mitigate ethical concerns, follow the guidelines mentioned here.

This brings us to an important point about the **ethics of web scraping**. While it's not an uncommon practice in the industry, and while it can result in very valuable data, it should not be done without considerations. For instance, you should be aware that web scraping can cause businesses to incur costs on your behalf, and that by requesting many pages in a short period of time, you are using up resources that were intended for actual users browsing the website. In order to mitigate these adverse consequences of web scraping, you should do the following:

- Limit the rate at which you make requests.

  For example, you might sleep the script for at least a few seconds between successive HTTP requests.

- Use a descriptive user agent to identify yourself to the website.

  For example, the default user agent for Python's **requests** library would be **python-requests/2.22.0** (for version 2.22.0 of **requests**).

  Check for a **robots.txt** file, which would live in the home folder of the website (for example, **www.website.com/robots.txt**). Adhere to the rules of this file by not scraping any page listed under a Disallow wildcard for relevant user agents.

While web scraping can have negative consequences if done unethically, the ability to programmatically request and parse information from websites can have very positive consequences as well. For example, this is how search engines such as Google are able to index websites, making them accessible via search. Furthermore, web scraping is very important for testing websites to make sure that things are functioning as expected, and that users are seeing the appropriate information across the site.

While some of these concepts may seem abstract right now, such as making HTTP requests or using web APIs, they will become clear as we continue to discuss them in the section that follows. In particular, you may learn best by seeing them done with Python, which is what we'll be doing in the exercises and activity to follow.

INTRODUCTION TO HTTP REQUESTS

The **Hypertext Transfer Protocol**, or **HTTP** for short, is the foundation of data communication for the internet. It defines how a page should be requested and how the response should look. For example, a client can request an Amazon page of laptops for sale, a Google search of local restaurants, or their Facebook feed. Along with the URL, the request will contain the user agent and available browsing cookies among the contents of the request header.

The user agent tells the server what browser and device the client is using, which is usually used to provide the most user-friendly version of the web page's response. Perhaps they have recently logged in to the web page; such information would be stored in a cookie that might be used to automatically log the user in.

These details of HTTP requests and responses are taken care of under the hood thanks to web browsers. Luckily for us, today, the same is true when making requests with high-level languages such as Python. For many purposes, the contents of request headers can be largely ignored. Unless specified otherwise, these are automatically generated in Python when requesting a URL. Still, for the purposes of troubleshooting and understanding the responses yielded by our requests, it's useful to have a foundational understanding of HTTP.

There are many types of HTTP methods, such as **GET**, **HEAD**, **POST**, and **PUT**. The first two are used for requesting that data be sent from the server to the client, whereas the last two are used for sending data to the server.

These HTTP methods can be summarized as follows:

- **GET**: Retrieves the information from the specified URL

- **HEAD**: Retrieves the meta-information from the HTTP header of the specified URL

- **POST**: Sends the attached information for appending to the resource(s) at the specified URL

- **PUT**: Sends the attached information for replacing the resource(s) at the specified URL

A **GET** request is sent each time we type a web page address into our browser and press *Enter*. For web scraping, this is usually the only HTTP method we are interested in, and it's the only method we'll be using in this chapter.

Once the request has been sent, a variety of response types can be returned from the server. These are labeled with 100-level to 500-level codes, where the first digit in the code represents the response class. These can be described as follows:

- **1xx**: An informational response; for example, the server is processing a request. It's uncommon to see this.

- **2xx**: Success; for example, the page has loaded properly.

- **3xx**: Redirection; for example, the requested resource has been moved and we were redirected to a new URL.

- **4xx**: Client error; for example, the requested resource does not exist.

- **5xx**: Server error; for example, the website server is receiving too much traffic and could not fulfill the request.

For the purposes of web scraping, we usually only care about the response class, that is, the first digit of the response code. However, subcategories of responses within each class exist that offer more granularity regarding what's going on. For example, a **401** code indicates an unauthorized response, whereas a **404** code indicates a page not found response. This distinction is noteworthy because a **404** would indicate we've requested a page that does not exist, whereas **401** tells us we need to log in to view the particular resource.

In the following exercise, we'll see how HTTP requests can be done in Python using Jupyter Notebooks.

MAKING HTTP REQUESTS WITH PYTHON

Now that we've talked about how HTTP requests work and what type of responses we should expect, let's see how this can be done in Python. We'll use a library called **requests**, which is one of the most popular (if not the most popular) external libraries for Python. Instead of using **requests**, it would also be possible to use Python's built-in tools, such as **urllib**, for making HTTP requests, but **requests** is far more intuitive and generally the best choice, as long as you're willing to add it as a dependency for your project.

requests allows for all sorts of customization with respect to headers, cookies, and authorization. It tracks redirects and provides methods for returning specific page content such as JSON (as we'll see in the API exercise, *Exercise 6.02*, *Making API calls with Python and Jupyter Notebooks*, later on).

One downside of **requests** is that it does not render JavaScript on the client-side (that is, on your machine). When requesting pages, servers will usually return HTML to you that have JavaScript code snippets included. If you're using a web browser (Chrome, Firefox, and so on), these snippets are automatically run on your machine during page load. When requesting content with Python using **requests**, however, this JavaScript code will not be executed. Therefore, any elements that would be altered or created by doing so will be missing.

Often, the lack of JavaScript rendering will not affect the ability to get the desired information from a page. If you find that rendering JavaScript is required for your use case, then you could consider doing this with a library such as Selenium. It has a similar API to the **requests** library but provides support for rendering JavaScript using web drivers. It can even run JavaScript commands on live pages, for example, to change the text color or scroll to the bottom of the page. As you could imagine, it is very useful for website testing purposes.

Let's dive into an exercise all about using the **requests** library with Python in a Jupyter Notebook.

EXERCISE 6.01: USING PYTHON AND JUPYTER NOTEBOOKS TO MAKE HTTP REQUESTS

Having learned the theory of HTTP requests, let's apply this in Python with Jupyter Notebooks. In this exercise, you will learn how to make HTTP requests in the notebook and how to interpret and work with the HTML response data. Follow these steps to complete this exercise:

1. Create a Jupyter Notebook and load the following libraries:

```
import pandas as pd
import numpy as np
import datetime
import time
import os

import matplotlib.pyplot as plt
%matplotlib inline
import seaborn as sns

%config InlineBackend.figure_format='retina'
sns.set() # Revert to matplotlib defaults
plt.rcParams['figure.figsize'] = (9, 6)
plt.rcParams['axes.labelpad'] = 10
sns.set_style("darkgrid")

%load_ext watermark
%watermark -d -v -m -p \
requests,numpy,pandas,matplotlib,seaborn,sklearn
```

2. Start by showing you a few ways to make HTTP requests with Python. Import the **requests** library, as follows:

```
import requests
```

3. Then, prepare a request by running the following code:

```
url = 'https://jupyter.org/'
req = requests.Request('GET', url)
req = req.prepare()
```

You use the **Request** class to prepare a **GET** request to the **jupyter.org** home page.

4. Print the docstring for the **req** prepared request by running **req?** in the next cell. You will get the following output:

```
Type:           PreparedRequest
String form:    <PreparedRequest [GET]>
File:           /anaconda3/lib/python3.7/site-packages/requests/models.py
Docstring:
The fully mutable :class:`PreparedRequest <PreparedRequest>` object,
containing the exact bytes that will be sent to the server.

Generated from either a :class:`Request <Request>` object or manually.

Usage::

  >>> import requests
  >>> req = requests.Request('GET', 'https://httpbin.org/get')
  >>> r = req.prepare()
  <PreparedRequest [GET]>

  >>> s = requests.Session()
  >>> s.send(r)
  <Response [200]>
```

Figure 6.1: Printing the docstring for a PreparedRequest object

Looking at its usage, you can see how the request can be sent using a session. This is similar to opening a web browser (starting a session) and then requesting a URL.

5. Make the request and store the response in a variable named **resp** by running the following code:

```
with requests.Session() as sess:
    resp = sess.send(req)
```

The preceding code returns the HTTP response, as referenced by the page variable. By using the **with** statement, you initialize a session whose scope is limited to the indented code block. This means you do not have to worry about explicitly closing the session, as this is done automatically.

6. Run the **resp** and **resp.status_code** commands to investigate the response. The string representation of the page should indicate a **200** status code response.

The output of the **resp** command is as follows:

```
<Response [200]>
```

The output of the **resp.status_code** command is **200**.

7. Assign the response text to the **page_html** variable and take a look at the first 1,000 characters of the string with the following command:

```
page_html = resp.text
page_html[:1000]
```

Here's the output showing first 100 characters:

```
'<!DOCTYPE html>\n<html>\n\n  <head>\n\n    <meta charset="utf-8">\n    <meta http-equiv="X-UA-Compa
tible" content="IE=edge">\n    <meta name="viewport" content="width=device-width, initial-scale=1">
\n    <meta name="description" content="">\n    <meta name="author" content="">\n\n    <title>Projec
t Jupyter | Home</title>\n    <meta property="og:title" content="Project Jupyter" />\n    <meta prop
erty="og:description" content="The Jupyter Notebook is a web-based interactive computing platform. T
he notebook combines live code, equations, narrative text, visualizations, interactive dashboards an
d other media.\n">\n    <meta property="og:url" content="https://www.jupyter.org" />\n    <meta prop
erty="og:image" content="https://jupyter.org/assets/homepage.png" />\n    <!-- Bootstrap Core CSS --
>\n    <link rel="stylesheet" href="/css/bootstrap.min.css">\n    <link rel="stylesheet" href="/css/
logo-nav.css?1580238146715752222">\n    <link rel="stylesheet" href="/css/cardlist.css">\n    <link
rel="stylesheet" href="'
```

Figure 6.2: Printing the response HTML content

As expected, the response is HTML.

You can format this output better with the help of **BeautifulSoup**, a library that will be used extensively for HTML parsing later in this section.

8. Print the **head** of the formatted HTML by running the following code:

```
from bs4 import BeautifulSoup
print(BeautifulSoup(page_html, 'html.parser').prettify()[:1000])
```

This displays the following output:

```
<!DOCTYPE html>
<html>
 <head>
  <meta charset="utf-8"/>
  <meta content="IE=edge" http-equiv="X-UA-Compatible"/>
  <meta content="width=device-width, initial-scale=1" name="viewport"/>
  <meta content="" name="description"/>
  <meta content="" name="author"/>
  <title>
   Project Jupyter | Home
  </title>
  <meta content="Project Jupyter" property="og:title">
   <meta content="The Jupyter Notebook is a web-based interactive computing platform. The notebook c
ombines live code, equations, narrative text, visualizations, interactive dashboards and other medi
a.
" property="og:description"/>
   <meta content="https://www.jupyter.org" property="og:url">
    <meta content="https://jupyter.org/assets/homepage.png" property="og:image">
    <!-- Bootstrap Core CSS -->
    <link href="/css/bootstrap.min.css" rel="stylesheet"/>
    <link href="/css/logo-nav.css?1580238146715752222" rel="stylesheet"/>
    <link href="/css/cardlist.css" rel="stylesheet"/>
    <link href="/css/github-buttons.cs
```

Figure 6.3: Displaying the response HTML content with indentation

You import **BeautifulSoup** and then print the output, where newlines are indented depending on their hierarchy in the HTML structure.

9. Take this a step further and actually display the HTML in Jupyter by using the IPython **display** module. Do this by running the following code:

```
from IPython.display import HTML
HTML(page_html)
```

Here's the screenshot of the output of this code:

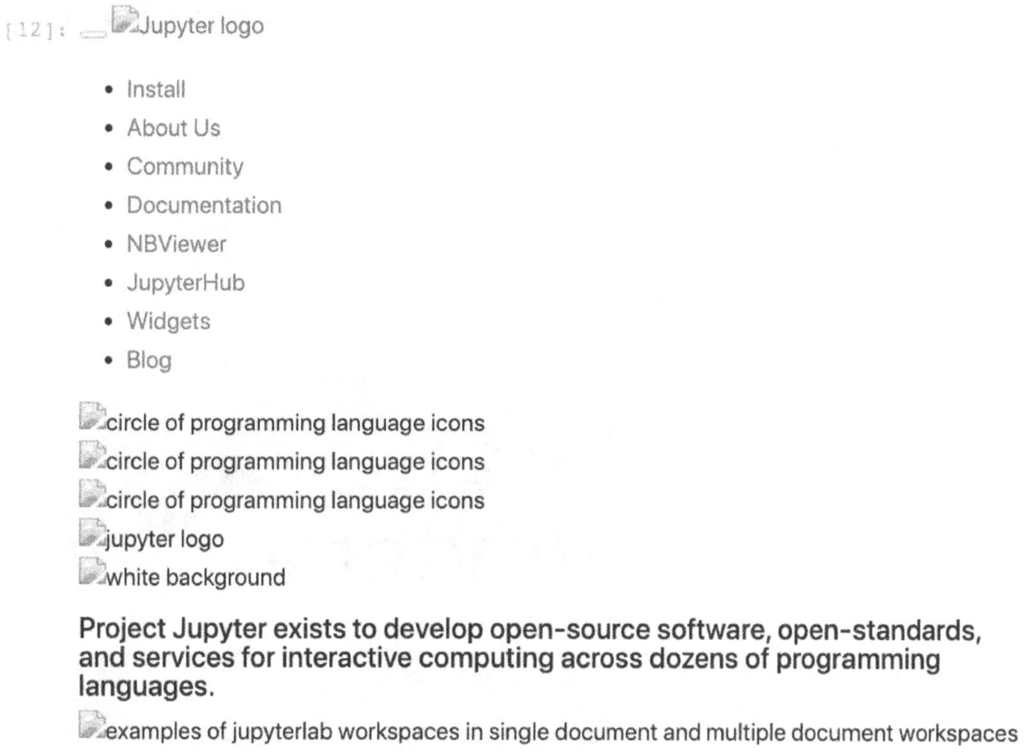

[12]: Jupyter logo

- Install
- About Us
- Community
- Documentation
- NBViewer
- JupyterHub
- Widgets
- Blog

circle of programming language icons
circle of programming language icons
circle of programming language icons
jupyter logo
white background

Project Jupyter exists to develop open-source software, open-standards, and services for interactive computing across dozens of programming languages.

examples of jupyterlab workspaces in single document and multiple document workspaces

Figure 6.4: HTML rendering without fetching images or running JavaScript

Here, you can see the HTML rendered as well as possible, given that no JavaScript code has been run and no external resources have been loaded. For example, the images that are hosted on the **jupyter.org** server are not rendered. Instead, we can see the alternate text—that is, **circle of programming icons**, **jupyter logo**, and so on.

10. Compare this to the live website, which can be opened in Jupyter using an IFrame, by running the following code:

```
from IPython.display import IFrame
IFrame(src=url, height=800, width=800)
```

Here's the screenshot of the IFrame:

Figure 6.5: Loading a live web page in the Jupyter Notebook

Here, you can see the full site rendered, including JavaScript and external resources. In fact, you can even click on the hyperlinks and load those pages in the IFrame, just like a regular browsing session.

> **NOTE**
>
> It's good practice to close the IFrame after using it. This prevents it from eating up memory and processing power. It can be closed by selecting the cell and clicking the **Current Outputs | Clear from the Cell** menu in the Jupyter Notebook.

11. At the start of this exercise, you made a request by preparing it and then used a session to send it. This is often done using a shorthand method instead, as seen here.

    Make a request to http://www.python.org/ by running the following code:

    ```
    url = 'http://www.python.org/'
    resp = requests.get(url)
    resp
    ```

    This will output the string representation of the page:

    ```
    <Response [200]>
    ```

    It should show a **200** status code, indicating a successful response to your request.

12. Run the following command to print the URL of your page:

    ```
    resp.url
    ```

    The output is as follows:

    ```
    'https://www.python.org/'
    ```

13. Run the following command to print the history attributes of the page.

    ```
    resp.history
    ```

    This will return **[<Response [301]>]**.

    > **NOTE**
    >
    > To access the source code for this specific section, please refer to https://packt.live/2ACHg63.
    >
    > You can also run this example online at https://packt.live/2zDrqYu.

The URL that's returned is not what we input; notice the difference? We were redirected from the input URL, http://www.python.org/, to the secured version of that page, https://www.python.org/. The difference is indicated by an additional **s** at the start of the URL, in the protocol. Any redirects are stored in the **history** attribute; in this case, we find one page in here with status code **301** (permanent redirect), corresponding to the original URL that was requested.

MAKING API CALLS WITH PYTHON

API calls can simply be an HTTP request, such as those that we looked at in the previous exercise. One difference is that API requests are generally expected to return data in machine-readable format, as opposed to *regular* web requests, which are expected to return HTML that browsers can render.

Another difference is that API calls are more likely to require authentication of some kind, such as passing a token parameter in the request URL or specifying request headers. Many APIs have more involved authentication methods, such as using the OAuth 2.0 protocol. These types of API requests are outside the scope of this book since we'll only be focusing on the simplest cases where no authentication is required.

In the next exercise, we'll show how to pull article information from Wikipedia using their free API. We'll use this API to extract data from a wiki table on interest rates by country. Then, in a later activity, *Activity 6.02*, *Analyzing Country Populations and Interest Rates*, we'll revisit this same table and learn how to extract the same data with HTML web scraping techniques.

EXERCISE 6.02: MAKING API CALLS WITH PYTHON AND JUPYTER NOTEBOOKS

API calls allow you to access well-structured data on demand. Knowing how to use them is a necessary skill for data scientists to have. In this exercise, you will work with the Wikipedia API in order to learn about how APIs can be used in general. You will make API requests and ingest the JSON response data. Follow these steps to complete this exercise:

1. Start up one of the following platforms for running Jupyter Notebooks:

   JupyterLab (run `jupyter lab`)

   Jupyter Notebook (run `jupyter notebook`)

2. Load the following libraries. You will use these to configure your plot settings for the Notebook:

```
import pandas as pd
import numpy as np
import datetime
import time
import os
```

```
import matplotlib.pyplot as plt
%matplotlib inline
import seaborn as sns
import requests
%config InlineBackend.figure_format='retina'
sns.set() # Revert to matplotlib defaults
plt.rcParams['figure.figsize'] = (9, 6)
plt.rcParams['axes.labelpad'] = 10
sns.set_style("darkgrid")

%load_ext watermark
%watermark -d -v -m -p \
requests,numpy,pandas,matplotlib,seaborn,sklearn
```

3. Run the following code to define your API request URL:

> **NOTE**
>
> Watch out for the slashes in the string below. Remember that the backslashes (\ ) are used to split the code across multiple lines, while the forward slashes (/) are part of the URL.

```
url = ('https://en.wikipedia.org/w/api.php' \
       '?action=parse' \
       '&page=List_of_countries_by_central_bank_interest_rates' \
       '&section=1' \
       '&prop=wikitext' \
       '&format=json')
url
```

This will result in the following output:

```
https://en.wikipedia.org/w/api.php?action=parse&page=List_
of_countries_by_central_bank_interest_
rates&section=1&prop=wikitext&format=json
```

Here, you are requesting the resource that satisfies a set of parameters, such as **action**, **page**, **section**, and so on. Notice that you have explicitly requested a response in **.json** format by appending **&format=json** to the URL. These parameters are specific to the Wikipedia API, but many APIs work in a similar way.

4. Make the API request by running the following code:

```
resp = requests.get(url)
resp
```

This will print the following output:

```
<Response [200]>
```

5. Run **resp.text[:100]** to print the first 100 lines of the response string. This should result in the following output:

```
'{"parse":{"title":"List of countries by central bank interest
rates","pageid":20582369,"wikitext":{"'
```

Notice how the string appears to represent JSON data, which is what we asked for when making the request.

6. Convert the string into a Python dictionary by running the following code:

```
data = resp.json()
type(data)
```

This should output the data object type, as follows:

```
dict
```

7. Run the **data** command to print the data object. Take note of some of the nested fields in the data, such as **parse**, **pageid**, and **wikitext**:

```
{'parse': {'title': 'List of countries by central bank interest rates',
  'pageid': 20582369,
  'wikitext': {'*': '== List ==\n{| class="wikitable sortable" style="text-align: center;"\n|- bgcol
or="#ececec" valign=bottom\n! Country or<br>currency union !! Central bank <br> interest rate (%) !!
```

Figure 6.6: The response data in JSON format

8. Extract the page title from the API response data by running the following command:

```
data['parse']['title']
```

This should output the following:

```
'List of countries by central bank interest rates'
```

9. Extract a row from the table contained in the API response data. This can be done by running the following code:

```
row_idx = 16


wikitext = data['parse']['wikitext']['*']
table_row = wikitext.split('|-')[row_idx]
table_row
```

This should output something similar to the following:

```
'                    \n|align="left"| {{flag|Canada}} || 1.75 ||
{{dts|format=dmy|2018-10-24}}<ref name="CentralBankNews"/><ref
name="GlobalRates">{{Cite web|url=http://www.global-rates.com/
interest-rates/central-banks/central-banks.aspx|title=Central banks -
summary of current interest rates|work=global-rates.com|accessdate=13
July 2017}}</ref>\n|1.40\n|0.35\n|1.25\n'
```

Ideally, the table data returned from Wikipedia's free API would be in a nicer format for us to ingest programmatically. As you can see, this is not quite the case. In the preceding output, you extracted the table from the response data as a **wikitext** string, and then separated the rows by splitting on **|-**.

10. Use regular expressions to parse data from the row. Get the country for the extracted row by running the following code:

```
import re
re.findall('flag\|([^}]+)}', table_row)
```

This should output the country name in a list, for example, the following:

```
['Canada']
```

Again, note that APIs would usually make this data easily available to the application using it. In this case, for Wikipedia, you are still able to access the data with relative ease by extracting the field in between **flag|** and **}**. In this case, you extracted **Canada** from **{{flag|Canada}}**.

11. Instead of using regular expressions, some data is easier to extract using Python string methods, such as **split** and **strip**. Get the interest rate for your extracted row by running the following command:

```
table_row.split('||')[1].strip()
```

This should output the interest rate as a string; for example:

```
'1.75'
```

By iterating over all of the rows in the API response data, you can apply this extraction to each and pull out all of the data for the requested table resource.

> **NOTE**
>
> To access the source code for this specific section, please refer to https://packt.live/2ACHg63.
>
> You can also run this example online at https://packt.live/2zDrqYu.

This concludes our section on API calls. In the exercise to follow, we'll be looking at the same Wikipedia data we used here and show how it can be extracted by parsing HTML, as opposed to extracting data by using an API call. This is purely for demonstration purposes, and you should always prefer using APIs as a data source, when available, compared to getting data from HTML scraping.

PARSING HTML WITH JUPYTER NOTEBOOKS

Scraping data from a web page involves making an HTTP request for the HTML resource, and then extracting data from the response content. An easy way to do this is by feeding this response content (HTML) into a high-level parsing library such as Python's **BeautifulSoup**. This is not to say this is the only way of doing this; in principle, it would be possible to pick out the data using regular expressions or Python string methods such as **split**. However, pursuing either of these options would be an inefficient use of time and could easily lead to errors. Therefore, it's generally frowned upon and, instead, the use of a trustworthy parsing tool is recommended.

In order to understand how content can be extracted from HTML, it's important to know the fundamentals of HTML. For starters, **HTML** stands for **Hyper Text Markup Language**. Like **Markdown** or **XML** (**eXtensible Markup Language**), it's simply a language for marking up text.

In HTML, the display text is contained within the content section of HTML elements; for example:

```
<p>Here is the text to display!</p>
```

In this piece of HTML, the content text to display is wrapped by **<p>** tags. Some common tag types are as follows:

- **<p>** (paragraph)

- **<div>** (text block)

- **<table>** (data table)

- **<h1>** (heading)

- **<img>** (image)

- **<a>** (hyperlinks)

Tags can have attributes, which specify important metadata. Most commonly, this metadata is used to control how and where the element text should appear on the page. This is where CSS files come into play. Consider the following example:

```
<p id="my-paragraph">Here is the text to display!</p>
```

In this piece of HTML, we are assigning an **id** to the **<p>** tag. This **id** can be referenced in a CSS file in order to set the style properties of the tag.

Attributes can store other useful information, such as the **href** hyperlink in an **<a>** tag, which specifies a URL link, or the alternate **alt** label in an **<img>** tag, which specifies the text to display if the image resource cannot be loaded. Consider the following example:

```
<a href="/my_picture_full_resolution.png">
<img src="/my_picture.png" alt="A photo of me!"></img>
</a>
```

In this piece of HTML, we're displaying an image that will be sourced from the **my_picture.png** resource. If this resource is not found, then the **alt** attribute text will be seen instead. This **<img>** element is then wrapped in an **<a>** tag, where the **href** attribute points to the location of the full-quality image. This will allow the user to navigate to that page by clicking on the image.

Now that we're properly equipped with some fundamental knowledge of HTML, let's turn our attention back to the Jupyter Notebook and parse some data. We'll also be using the developer tools window from the Chrome web browser, since it's very helpful when it comes to HTML parsing.

EXERCISE 6.03: PARSING HTML WITH PYTHON AND JUPYTER NOTEBOOKS

In this exercise, we'll focus on extracting data from HTML documents. You will learn how to implement Python parsing techniques and see why Jupyter Notebooks are such a good fit for this task. Follow these steps to complete this exercise:

1. Create a new notebook using either of the following commands:

   JupyterLab (run **jupyter lab**)

   Jupyter Notebook (run **jupyter notebook**)

2. Run the following code to load some libraries. You will use these libraries to configure your plot settings for the notebook:

```
import pandas as pd
import numpy as np
import datetime
import time
import os

import matplotlib.pyplot as plt
%matplotlib inline
import seaborn as sns
import requests

%config InlineBackend.figure_format='retina'
sns.set() # Revert to matplotlib defaults
plt.rcParams['figure.figsize'] = (9, 6)
plt.rcParams['axes.labelpad'] = 10
sns.set_style("darkgrid")

%load_ext watermark
%watermark -d -v -m -p \
requests,numpy,pandas,matplotlib,seaborn,sklearn
```

3. Scrape the central bank interest rates for each country, as reported by Wikipedia. Before diving into the code, open up the web page containing the data to be extracted.

    Go to https://en.wikipedia.org/wiki/List_of_countries_by_central_bank_interest_rates in a web browser. Use Chrome, if possible, as later in this exercise, we'll show you how to view and search the HTML using Chrome's developer tools.

    Looking at the page, you can see very little content other than a big list of countries and their interest rates. This is the table you'll be scraping.

4. Return to the Jupyter Notebook and request the page by running the following code:

```
url = 'https://en.wikipedia.org/wiki/List_of_countries_by_'\
      'central_bank_interest_rates'
resp = requests.get(url)
print(resp.url, resp.status_code)
```

    This should output the URL, followed by a **200** status code indicating success:

```
https://en.wikipedia.org/wiki/List_of_countries_by_central_bank_
interest_rates 200
```

5. Load the HTML as a **BeautifulSoup** object so that it can be parsed. Do this by running the following code:

```
from bs4 import BeautifulSoup
soup = BeautifulSoup(resp.content, 'html.parser')
```

> **NOTE**
>
> We're using Python's default **html.parser** as the parser, but other parsing libraries, such as **lxml**, can be installed and used instead. Since each HTML parser has different logic for interpreting documents, the resulting **BeautifulSoup** object may vary, depending on which is used.

6. Usually, when working with a new object in Jupyter, such as the
   BeautifulSoup object you created previously, it's a good idea to pull up
   the docstring.

   Do this by running **soup?**, as shown in the following screenshot:

```
Signature:      soup(*args, **kwargs)
Type:           BeautifulSoup
String form:
<!DOCTYPE html>

            <html class="client-nojs" dir="ltr" lang="en">
            <head>
            <meta charset="utf-8"/>
            <t <...> (){mw.config.set({"wgBackendResponseTime":137,"wgHostname":"mw1248"});});</scrip
></body></html>

Length:         4
File:           /anaconda3/lib/python3.7/site-packages/bs4/__init__.py
Docstring:
This class defines the basic interface called by the tree builders.

These methods will be called by the parser:
  reset()
  feed(markup)

The tree builder may call these methods from its feed() implementation:
  handle_starttag(name, attrs) # See note about return value
  handle_endtag(name)
  handle_data(data) # Appends to the current data node
  endData(containerClass) # Ends the current data node
```

Figure 6.7: The docstring for a BeautifulSoup object

As can be seen, it's not particularly helpful in this case because the docstring is not very informative.

Another tool for exploring Python objects is the built-in **dir** function, which lists the attributes and methods of an object. Call this by running **dir(soup)**, as shown here:

```
[30]: ['ASCII_SPACES',
       'DEFAULT_BUILDER_FEATURES',
       'NO_PARSER_SPECIFIED_WARNING',
       'ROOT_TAG_NAME',
       '__bool__',
       '__call__',
       '__class__',
       '__contains__',
       '__copy__',
       '__delattr__',
       '__delitem__',
       '__dict__',
       '__dir__',
       '__doc__',
       '__eq__',
       '__format__',
       '__ge__',
       '__getattr__',
       '__getattribute__',
```

Figure 6.8: The output of dir(soup)

Scrolling through the list, you will see the methods and attributes that we'll be using later, such as **find_all**, **attrs**, and **text**. Still, this is not particularly informative.

7. There is yet another way of getting information on Python objects, which you will see here. Install the external library called **pdir2** with **pip**, by running the following in your Terminal:

```
pip install pdir2
```

Once installed, this can be used by running the following code:

```
import pdir
pdir(soup)
```

Notice that we import **pdir**, even though the package is listed on the **Python Packaging Index (PyPI)** as **pdir2**:

```
[31]: property:
          ASCII_SPACES, DEFAULT_BUILDER_FEATURES, NO_PARSER_SPECIFIED_WARNING, ROOT_TAG_NAME, _most_recent_element, _namespaces, attr
      s, builder, can_be_empty_element, cdata_list_attributes, contains_replacement_characters, contents, currentTag, current_data, d
      eclared_html_encoding, element_classes, hidden, is_xml, known_xml, markup, name, namespace, next_element, next_sibling, origina
      l_encoding, parent, parse_only, prefix, preserve_whitespace_tag_stack, preserve_whitespace_tags, previous_element, previous_sib
      ling, tagStack
      special attribute:
          __class__, __dict__, __doc__, __module__, __weakref__
      abstract class:
          __subclasshook__
      object customization:
          __bool__, __format__, __hash__, __init__, __new__, __repr__, __sizeof__, __str__
      rich comparison:
          __eq__, __ge__, __gt__, __le__, __lt__, __ne__
      attribute access:
          __delattr__, __dir__, __getattr__, __getattribute__, __setattr__
      class customization:
          __init_subclass__
      container:
          __contains__, __delitem__, __getitem__, __iter__, __len__, __setitem__
      copy:
          __copy__
      pickle:
          __getstate__, __reduce__, __reduce_ex__
      descriptor:
          _is_xml: @property with getter, Is this element part of an XML tree or an HTML tree?
          children: @property with getter
          descendants: @property with getter
          isSelfClosing: @property with getter, Is this tag an empty-element tag? (aka a self-closing tag)
          is_empty_element: @property with getter, Is this tag an empty-element tag? (aka a self-closing tag)
          next: @property with getter
          nextSibling: @property with getter, setter
          next_elements: @property with getter
          next_siblings: @property with getter
          parents: @property with getter
          parserClass: @property with getter, setter
          previous: @property with getter
```

Figure 6.9: The output of pdir(soup)

Here, you can see a similar list of methods and attributes that can be called on soup, but now they are organized into groupings, and descriptions are included where applicable.

Since we'll be using the **find_all** method, let's search for that description in the list. It should read as follows:

```
findPrevious: Returns the first item that matches the given criteria and
findPreviousSibling: Returns the closest sibling to this Tag that matches the
                                                            ch the given
find_all: Extracts a list of Tag objects that match the given
                                                          and appear
find_all_previous: Returns all items that match the given criteria and appear
find_next: Returns the first item that matches the given criteria and
```

Figure 6.10: Description of the find_all method

8. It's time to start parsing data from our HTML. To start with, get the **h1** heading for the page by running the following code:

```
h1 = soup.find_all('h1')
h1
```

This should output a list that contains the page title element:

```
[<h1 class="firstHeading" id="firstHeading" lang="en">List of countries
by central bank interest rates</h1>]
```

Usually, pages only have one **h1** (top-level heading) element, so it's no surprise that there is only one found here.

9. At this point, you have identified the HTML element that contains your data, but the field still needs to be extracted as a string. To do this, run the next few cells one by one in the notebook:

```
h1 = h1[0]
```

Print the HTML element attributes as follows:

```
h1.attrs
```

The output is as follows:

```
{'id': 'firstHeading', 'class': ['firstHeading'], 'lang': 'en'}
```

Print the visible text as follows

```
h1.text
```

The output is as follows:

```
'List of countries by central bank interest rates'
```

The following screenshot shows the output of the preceding code:

First, you assign the **h1** variable to the first (and only) list element with **h1 = h1[0]**.

Then, print out the HTML element attributes with **h1.attrs**. Here, you can see the **id** and **class** elements, both of which can be referenced in CSS stylesheets.

Finally, you get the title as plain text by running **h1.text**.

10. Run the following to see the number of image tags you were able to extract:

```
imgs = soup.find_all('img')
len(imgs)
```

This should return a number around 100, indicating the number of images on the page.

11. In this case, most of these images correspond to country flags in the table. This can be seen by printing the source of each image. Do this by running the following code:

```
for element in imgs:
    if 'src' in element.attrs.keys():
        print(element.attrs['src'])
```

This should output the path of each image resource, as shown in the following sample:

```
//upload.wikimedia.org/wikipedia/commons/thumb/3/36/Flag_of_Albania.
svg/21px-Flag_of_Albania.svg.png
//upload.wikimedia.org/wikipedia/commons/thumb/9/9d/Flag_of_Angola.
svg/23px-Flag_of_Angola.svg.png
//upload.wikimedia.org/wikipedia/commons/thumb/1/1a/Flag_of_
Argentina.svg/23px-Flag_of_Argentina.svg.png
```

12. Now, scrape the data from the table. Use Chrome's developer tools to hunt down the relevant HTML elements.

   If you haven't done so already, open the Wikipedia page in Chrome. Then, in the browser, select **Developer Tools** from the **More Tools** menu button, which is shown with three vertical dots:

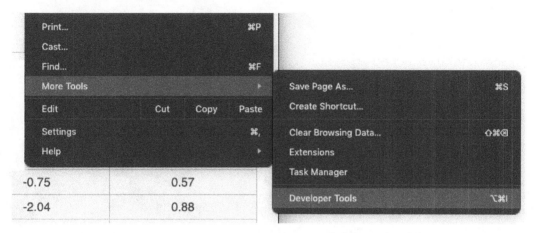

Figure 6.11: Opening the developer tools menu

   You can also open the **Developer Tools** by pressing *Ctrl + Shift + I* on Windows or Linux, or *Ctrl + Option + I* on Mac. A sidebar will open where you can view the HTML. You can do this by clicking on the **Elements** tab.

13. Select the arrow icon at the top left of the tools sidebar. This allows us to hover over the page and see where the HTML element is located. Do this by going to the **Elements** section of the sidebar:

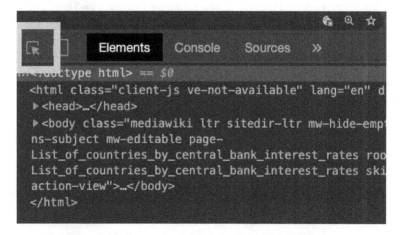

Figure 6.12: Arrow icon for locating HTML elements

14. Hover over the body to see how the table is contained within the **div** that has **id="bodyContent"**.

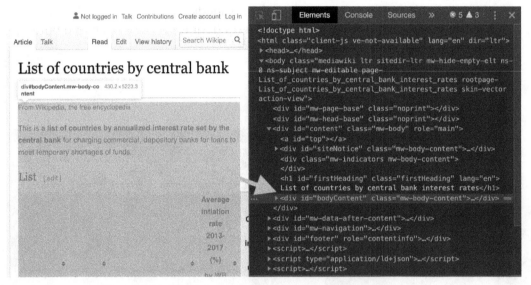

Figure 6.13: The parent div of the table we want to extract

15. Now, go back to your notebook and select that **div** by running the following command:

```
body_content = soup.find('div', {'id': 'bodyContent'})
```

Here, you're using the **find** method, which is identical to the **find_all** method you used previously, except that it returns only the first match. When calling this, you passed a second argument, **{'id': 'bodyContent'}**, which follows the form **{attribute_name: attribute_value}**.

16. Having narrowed down the content of interest, continue to seek out the table within this subset of the full HTML.

Usually, tables are organized into headers (**<th>**), rows (**<tr>**), and data entries (**<td>**). Using this knowledge, attempt to get the table headers by running the following code:

```
table_headers = body_content.find_all('th')
table_headers
```

This should output a list of heading elements, starting with those shown here:

```
[<th>Country or<br/>currency union</th>,
 <th>Central bank <br/> interest rate (%)</th>,
```

```
<th>Date of last <br/> change
</th>,
<th>Average inflation rate 2013-2017 (%)
...
```

17. Our next step is parsing the headers as plain text, from this list of HTML elements. Do this by running the following code:

```
for i, t in enumerate(table_headers):
    print(i, t.text.strip())
    print('-'*10)
```

This should output the column names, starting with those shown here:

```
0 Country orcurrency union
----------
1 Central bank  interest rate (%)
----------
2 Date of last  change
----------
3 Average inflation rate 2013-2017 (%)
by WB and IMF[1][2] as in the List
----------
...
```

18. You will now extract data for the first four columns of the table. Using the preceding output as a reference, manually set the header names for our table by running the following code:

```
table_headers = ['Country or currency union', \
                 'Central bank interest rate (%)', \
                 'Date of last change', \
                 'Average inflation rate (%)']
```

19. Now, you are ready to extract the data., first figuring out how this should be done on a per-row basis.

    Select the HTML element for a row in the table by running the following code:

```
row_number = 8
row_data = body_content.find_all('tr')[row_number]\
           .find_all('td')
```

Whereas you searched for all the header elements before, here, you are looking for all the rows and then selecting them at index 8 (selected at random). Then, in the same line of code, search for all the data elements, **td**, in that row.

20. Run the **row_data** command in the notebook to see the resulting data elements that make up the row:

```
[<td align="left"><span class="flagicon"><img alt="" class="thumbborder" data-file-height="900" data-
file-width="1500" decoding="async" height="14" src="//upload.wikimedia.org/wikipedia/commons/thumb/2/
2c/Flag_of_Bahrain.svg/23px-Flag_of_Bahrain.svg.png" srcset="//upload.wikimedia.org/wikipedia/common
s/thumb/2/2c/Flag_of_Bahrain.svg/35px-Flag_of_Bahrain.svg.png 1.5x, //upload.wikimedia.org/wikipedia/
commons/thumb/2/2c/Flag_of_Bahrain.svg/46px-Flag_of_Bahrain.svg.png 2x" width="23"/> </span><a href
="/wiki/Bahrain" title="Bahrain">Bahrain</a></td>,
 <td>2.50</td>,
 <td><span data-sort-value="000000002019-07-31-0000" style="white-space:nowrap">31 July 2019</span><s
up class="reference" id="cite_ref-CentralBankNews_3-6"><a href="#cite_note-CentralBankNews-3">[3]</a>
</sup>
 </td>,
 <td>2.40
 </td>,
 <td>0.10
 </td>,
 <td>1.04
 </td>]
```

Figure 6.14: The data elements in a selected row

21. Recall when you iterated through the header elements and pulled the text for each. Here, you will do an analogous operation on the row data.

Run the following code:

```
for i, d in enumerate(row_data):
    print(i, d.text)
```

This should output the row data and indices so that they are easy to interpret, as follows:

```
0  Bahrain
1 2.50
2 31 July 2019[3]


3 2.40


4 0.10


5 1.04
```

Using this row as an example, look at each element of interest and determine how to best parse the data entry. Recall that you are interested in the following columns: **Country or currency union**, **Central bank interest rate (%)**, **Date of last change**, and **Average inflation rate (%)**.

22. The first entry you are interested in is the country. Assign the **d1** variable to the HTML element for that entry by running the following code:

```
d1 = row_data[0]
d1
```

This should output the unparsed element as a string, as shown here:

```
<td align="left"><span class="flagicon"><img alt=""
class="thumbborder" data-file-height="900" data-file-width="1500"
decoding="async" height="14" src="//upload.wikimedia.org/wikipedia/
commons/thumb/2/2c/Flag_of_Bahrain.svg/23px-Flag_of_Bahrain.svg.png"
srcset="//upload.wikimedia.org/wikipedia/commons/thumb/2/2c/Flag_of_
Bahrain.svg/35px-Flag_of_Bahrain.svg.png 1.5x, //upload.wikimedia.
org/wikipedia/commons/thumb/2/2c/Flag_of_Bahrain.svg/46px-Flag_of_
Bahrain.svg.png 2x" width="23"/> </span><a href="/wiki/Bahrain"
title="Bahrain">Bahrain</a></td>
```

By looking at the preceding element, you can see that our country name is embedded in an **<a>** tag. Therefore, the country can be cleanly parsed by searching for that **<a>** tag, and then getting the text content.

23. First, run **d1.text** and notice how the output has a character such as **\xa0** before the country. This is clearly not wanted.

24. Next, run the following three commands, where you will find the **<a>** tag and then get its content, which should be the clean country name, as expected:

```
d1.txt
d1.find('a')
d1.find('a').text
```

The **d1.txt** command will return **'\xa0Bahrain'**, **d1.find('a')** will return **<a href="/wiki/Bahrain" title="Bahrain">Bahrain</a>**, and **d1.find('a').text** command returns **'Bahrain'**.

25. Moving on to the second data element, which will correspond to the interest rate, run the next couple of cells in the notebook to parse the value:

```
d2 = row_data[1]
d2
```

This will return **<td>2.50</td>**.

Note how we leave the number as a string, for now:

```
d2.text
```

The output is **'2.50'**.

26. For the next data element, you need to parse the date. Look at the element itself by running the following code:

```
d3 = row_data[2]
d3
```

This should output the unparsed element as a string, as shown here:

```
<td><span data-sort-value="000000002019-05-24-0000" style="white-
space:nowrap">24 May 2019</span><sup class="reference" id="cite_
ref-CentralBankNews_3-1"><a href="#cite_note-CentralBankNews-3">[3]</
a></sup>
</td>
```

Similar to the country name entry, you can see that the date is embedded within another HTML element. In particular, it's in a **<span>** tag.

Like you did previously, first run **d3.text** and notice how the output has unwanted characters such as "[3]\n", which follow the date.

Next, run the **d3.find_all('span')** and **d3.find_all('span')[0].text** commands, where you find the **<span>** tag and then get its content, which should be the clean date, as expected:

```
[52]: d3.text
```

```
[52]: '31 July 2019[3]\n'
```

```
[53]: d3.find_all('span')
```

```
[53]: [<span data-sort-value="000000002019-07-31-0000" style="white-space:nowrap">31 J
      uly 2019</span>]
```

```
[54]: d3.find_all('span')[0].text
```

```
[54]: '31 July 2019'
```

Figure 6.15: Parsing the date from a span element

27. The final entry we are interested in is the inflation rate. Look at this element by running the following code:

```
d4 = row_data[3]
d4
```

This should output the unparsed element as a string, as shown here:

```
<td>2.40
</td>
```

28. Run the **d4.text** and **d4.text.strip()** commands to parse this data. Notice how, after running **d4.text**, we have an unwanted newline character. This can be removed by calling the **strip** string method.

Executing **d4.text** will return **'2.40\n'** and executing **d4.text.strip()** will return **'2.40'**.

29. Having written the proper code for parsing a row, you are ready to perform the full scrape. This is done by iterating over the row elements, **<th>**, and attempting to extract the data for each, like you did previously. Do this by running the following code:

```
int_rates_data = []
row_elements = body_content.find_all('tr')
for i, row in enumerate(row_elements):
    row_data = row.find_all('td')
    if len(row_data) < 3:
        print('Ignoring row {} because length < 4'.format(i))
        continue

    d1, d2, d3, d4 = row_data[:4]
    errs = []

    try:
        d1 = d1.find('a').text
    except Exception as e:
        d1 = ''
        errs.append(str(e))
```

```
try:
    d2 = d2.text
except Exception as e:
    d2 = ''
    errs.append(str(e))

try:
    d3 = d3.find_all('span')[0].text
except Exception as e:
    d3 = ''
    errs.append(str(e))

try:
    d4 = d4.text.strip()
except Exception as e:
    d4 = ''
    errs.append(str(e))

data = [d1, d2, d3, d4]
print(data)
int_rates_data.append(data)
if errs:
    print('Errors in row {}: {}'.format(i, ', '.join(errs)))
```

There are a few interesting parts of this loop:

- We watch for rows with unexpected qualities – in this case, ones that have a length of less than 4 – and ignore them.

- We use **try...except** logic to deal with rows that have errors.

- We print out data as we iterate.

- Errors do not go unnoticed! We print them out as well as we iterate.

The output should be similar to the following:

```
Ignoring row 0 because length < 4
['Albania', '1.00', '6 June 2016', '1.75']
['Angola', '15.50', '24 May 2019', '17.54']
['Argentina', '68.00', '15 October 2019', '31.17']
['Armenia', '5.75', '29 January 2019', '2.38']
['Australia', '0.75', '1 October 2019', '1.93']
['Azerbaijan', '8.25', '26 July 2019', '6.51']
['Bahamas', '4.00', '22 December 2016', '0.97']
...

Ignoring row 99 because length < 4
Ignoring row 100 because length < 4
Ignoring row 101 because length < 4
Ignoring row 102 because length < 4
Ignoring row 103 because length < 4
```

You will notice that some rows raised errors, such as the following message:

```
Errors in row 26: list index out of range
```

While this error affected **row 26** at the time of writing, this may change in the future. All you know from this message is that a list index was out of range, so let's try to understand this by looking at it in more detail.

30. Pick out the bad row by running the following command:

```
bad_row_26 = body_content.find_all('tr')[26]
```

If needed, please adjust the index accordingly to reflect the errors that were raised in parsing the data for your particular table.

31. Next, unpack the row entries into the same variables you used previously for figuring out how to parse our data. Do this by running the following command:

```
d1, d2, d3, d4 = bad_row_26.find_all('td')[:4]
```

32. Now, you can attempt to parse the data from each of these items by using the same methods as before, until you identify where the error came from. Do this by running the following few cells:

```
d1.find('a').text
float(d2.text)
d3.find_all('span')[0].text
```

Upon running these cells, you will see the following output:

```
[61]:  d1.find('a').text

[61]:  'Eastern Caribbean'

[62]:  float(d2.text)

[62]:  6.5

[63]:  d3.find_all('span')[0].text
```

```
---------------------------------------------------------------------------
IndexError                                Traceback (most recent call last)
<ipython-input-63-a4aac356ecf6> in <module>
----> 1 d3.find_all('span')[0].text

IndexError: list index out of range
```

Figure 6.16: Troubleshooting for a row that could not be parsed

As can be seen, the first and second entries parse properly, but the third entry fails to parse and returns the **index out of range** error that was seen initially.

33. Here, you can see that **find_all('span')** seems to have returned no match, and therefore fails to look up the first (position "0") index. Confirm that this is the case by running the following command:

```
d3.find_all('span')
```

This should output an empty list, as expected.

34. Now that you have parsed your data, write it to disk.

    First, print the table headers that we set earlier using **table_ headers** command:

```
['Country or currency union',
 'Central bank interest rate (%)',
 'Date of last change',
 'Average inflation rate (%)']
```

35. Next, print the data that we parsed from the page as follows:

```
int_rates_data
```

    The output is as follows:

```
[['Albania', '1.00', '6 June 2016', '1.75'],
 ['Angola', '15.50', '24 May 2019', '17.54'],
 ['Argentina', '68.00', '15 October 2019', '31.17'],
 ['Armenia', '5.75', '29 January 2019', '2.38'],
 ['Australia', '0.75', '1 October 2019', '1.93'],
 ['Azerbaijan', '8.25', '26 July 2019', '6.51'],
 ['Bahamas', '4.00', '22 December 2016', '0.97'],
 ['Bahrain', '2.50', '31 July 2019', '2.40'],
 ['Bangladesh', '6.00', '6 April 2018', '6.38'],
 ['Barbados', '7.00', '1 June 2009', '1.67'],
 ['Belarus', '9.50', '7 August 2019', '13.47'],
 ['Botswana', '5.00', '24 October 2017', '3.89'],
 ['Brazil', '4.25', '5 February 2020', '6.73'],
 ['Bulgaria', '0.00', '29 January 2016', '0.12'],
 ['Canada', '1.75', '24 October 2018', '1.40'],
```

Figure 6.17: The data that was extracted from Wikipedia

36. In order to write this data to disk, you are going to use the pandas library. Save the data into a CSV file by running the following code:

```
f_path = '../data/countries/interest_rates_raw.csv'
pd.DataFrame(int_rates_data, columns=table_headers)\
           .to_csv(f_path, index=False)
```

37. Check that the data has been written properly by opening the CSV file with a text reader or Excel. If your Jupyter environment supports bash, then you can check the **head** of the table by running the following code:

```
%%bash
head ../data/countries/interest_rates_raw.csv
```

The output displayed is shown here:

```
Country or currency union,Central bank interest rate (%),Date of last change,Av
erage inflation rate (%)
Albania,1.00,6 June 2016,1.75
Angola,15.50,24 May 2019,17.54
Argentina,68.00,15 October 2019,31.17
Armenia,5.75,29 January 2019,2.38
Australia,0.75,1 October 2019,1.93
Azerbaijan,8.25,26 July 2019,6.51
Bahamas,4.00,22 December 2016,0.97
Bahrain,2.50,31 July 2019,2.40
Bangladesh,6.00,6 April 2018,6.38
```

Figure 6.18: The first 10 rows of the extracted Wikipedia data in a CSV file

> **NOTE**
>
> To access the source code for this specific section, please refer to https://packt.live/2ACHg63.
>
> You can also run this example online at https://packt.live/2zDrqYu.

That's it for the first web scraping exercise. Hopefully, you enjoyed putting what you've learned about HTTP requests and HTML into practice by scraping interest rate data by country. In the following activity, you'll have the opportunity to attempt the same process on your own, in order to pull a table that lists the population of each country.

ACTIVITY 6.01: WEB SCRAPING WITH JUPYTER NOTEBOOK

In this activity, we are going to get the population of each country. Then, in the next topic, we'll combine this data with the data we pulled from the previous exercise, in order to create a dataset that can be used for analysis.

Here, we'll look at another Wikipedia page, which is available at https://en.wikipedia.org/wiki/List_of_countries_and_dependencies_by_population.

Our aim is to apply the concepts that we looked at previously to a new web page with different data.

In order to do this, follow these steps:

1. Create a new Jupyter Notebook and load the necessary libraries.

2. Run the following code to assign the **url** as a variable:

```
url = 'https://en.wikipedia.org/wiki/List_of_countries_and'\
      '_dependencies_by_population'
```

3. Render an IFrame in the notebook with a live version of the web page. Then, close this by selecting **Cell | Current Outputs | Clear**.

4. Use the **requests** library to request the page.

5. Instantiate a **BeautifulSoup** object, which we can use to parse the page HTML.

6. Find the **h1** of the page. of the page.

7. Select the **<div>** tag with **id:bodyContent**.

8. Search this **div** element for the table headers and print the element text for each, along with their index.

9. Manually set the table headers in a variable. We are interested in the columns with indices 1-4, as follows: **Country** (or dependent territory), **Population**, **% of World Population**, and **Date**.

10. Find all the **<tr>** elements (rows) of the table. Then, select one of them and find all the **<td>** elements (data entries) in that row.

11. Determine the number of data entries in the row.

12. Print out the data entry elements you found.

13. Iterate over the data entries and print their text, along with the index number.

14. Select the elements we are interested in parsing. Recall that this will correspond to indices 1-4.

15. Assign the **d1**, **d2**, **d3**, and **d4** variables to each index and run through the same process that we followed in the previous exercise in order to parse clean data entries for each. For now, leave any numeric field as a string.

16. Once you have figured out how to parse the data from each entry, iterate over the rows, just like we did in the previous exercise, and parse the relevant data for each.

17. Print out the data you were able to pull for the table. This should have four columns.

18. Print out the corresponding headers, all of which were manually set earlier in this activity.

19. Write the data into a CSV file using pandas. Use the `../data/countries/ populations_raw.csv` file path.

20. Open up your CSV file and make sure it looks as expected.

> **NOTE**
>
> The solution to this activity can be found on page 301.

To summarize, we've seen how Jupyter Notebooks can be used for web scraping. We started this chapter by learning about HTTP methods and status codes. Then, we used the **requests** library to actually perform HTTP requests with Python and saw how the **BeautifulSoup** library can be used to parse response HTML.

Our Jupyter Notebooks turned out to be great for this type of work. We were able to explore the results of our web requests and experiment with various HTML parsing techniques, render the HTML, and even load a live version of the web page inside the notebook!

In the next section of this chapter, we'll process the raw data we've collected thus far with pandas in order to prepare it for analysis. Then, we'll close out this chapter with an activity that ties the concepts that we learned about from earlier chapters together using our new web-scraped data.

DATA WORKFLOW WITH PANDAS

As we've seen time and time again in this book, pandas is an integral part of performing data science with Python and Jupyter Notebooks. DataFrames offer us a way to organize and store labeled data, but more importantly, pandas provides time-saving methods for transforming data. Examples we have seen in this book include dropping duplicates, mapping dictionaries to columns, applying functions over columns, and filling in missing values.

In the next exercise, we'll reload the raw tables that we pulled from Wikipedia, clean them up, and merge them together. This will result in a dataset that is suitable for analysis, which we'll use for a final exercise, where you'll have an opportunity to perform exploratory analysis and apply the modeling concepts that you learned about in earlier chapters.

EXERCISE 6.04: PROCESSING DATA FOR ANALYSIS WITH PANDAS

In this exercise, we continue working on the country data that was pulled from Wikipedia in the preceding sections. Recall that we extracted the central bank interest rates and populations of each country, and saved the raw results in CSV files. We'll load the data from these files and process them to prepare the datasets for analysis. This will involve renaming columns, dropping missing data, and making sure the datatypes of each column are appropriate. Follow these steps to complete this exercise:

1. Start up one of the following platforms for running Jupyter Notebooks:

    JupyterLab (run **jupyter lab**)

    Jupyter Notebook (run **jupyter notebook**)

2. Run the following code to load the libraries that are required for this exercise:

```
import pandas as pd
import numpy as np
import datetime
import time
import os

import matplotlib.pyplot as plt
%matplotlib inline
import seaborn as sns

%config InlineBackend.figure_format='retina'
```

```
sns.set() # Revert to matplotlib defaults
plt.rcParams['figure.figsize'] = (9, 6)
plt.rcParams['axes.labelpad'] = 10
sns.set_style("darkgrid")

%load_ext watermark
%watermark -d -v -m -p \
requests,numpy,pandas,matplotlib,seaborn,sklearn
```

3. Load the *raw* data that you pulled in the previous sections by running the following code:

```
df_populations = pd.read_csv('../data/countries'\
                             '/populations_raw.csv')
df_int_rates = (pd.read_csv('../data/countries'\
                            '/interest_rates_raw.csv'))
```

> **NOTE**
>
> This file can also be found in this book's GitHub repository at https://packt.live/2MYlCvs.

4. Run the following command to check the first five rows of **df_populations**:

```
df_populations.head()
```

The output is as follows:

	Country(or dependent territory)	Population	% of WorldPopulation	Date
0	NaN	Population	% of worldpopulation	Date
1	China	1,403,612,840	18.0%	20 Jul 2020
2	India	1,364,925,239	17.5%	20 Jul 2020
3	United States	329,985,664	4.23%	20 Jul 2020
4	Indonesia	269,603,400	3.46%	1 Jul 2020

Figure 6.19: First five rows of df_populations

5. Run the following command to check the head **df_int_rates**:

```
df_int_rates.head()
```

The output is as follows:

	Country or currency union	Central bank interest rate (%)	Date of last change	Average inflation rate (%)
0	Albania	1.00	6 June 2016	1.75
1	Angola	15.50	24 May 2019	17.54
2	Argentina	38.00	5 March 2020	31.17
3	Armenia	5.75	29 January 2019	2.38
4	Australia	0.25	19 March 2020	1.93

Figure 6.20: The head of each table we extracted from Wikipedia

You will now clean up each table by making sure each column has the appropriate datatype and looking for rows with missing entries. This will give you the opportunity to do exploratory analysis.

6. Set the display limits for pandas DataFrames so that you can see up to 10,000 rows from each table (although there are far less in these datasets). Do this by running the following command:

```
pd.options.display.max_rows = 10000
```

7. Print the entire **populations** table and scan down the rows in your notebook.

	Country(or dependent territory)	Population	% of WorldPopulation	Date
0	NaN	Population	% of worldpopulation	Date
1	China	1,403,612,840	18.0%	20 Jul 2020
2	India	1,364,925,239	17.5%	20 Jul 2020
3	United States	329,985,664	4.23%	20 Jul 2020
4	Indonesia	269,603,400	3.46%	1 Jul 2020
5	Pakistan	220,892,331	2.83%	1 Jul 2020
6	Brazil	211,817,730	2.72%	20 Jul 2020
7	Nigeria	206,139,587	2.64%	1 Jul 2020
8	Bangladesh	168,985,154	2.17%	20 Jul 2020
9	Russia	146,748,590	1.88%	1 Jan 2020

Figure 6.21: Displaying the full populations table in Jupyter

8. First, deal with the column names. Print these out by running the following command:

```
df_populations.columns
```

This should output the following:

```
Index(['Country(or dependent territory)', 'Population',
       '% of WorldPopulation', 'Date'],
      dtype='object')
```

9. When working with data in pandas, you want column names to be descriptive and easy to type. It is good practice to ensure lowercase words are separated by underscores, as per the Python naming convention.

Manually set the column names by running the following code:

```
df_populations.columns = ['country', 'population', \
                          'population_pct', 'date',]
```

10. Print the datatypes of each column by running **df_populations.dtypes**. This should output the following:

```
country           object
population        object
population_pct    object
date              object
dtype: object
```

Each column seems to be of the **object** type, which can be interpreted as a **string** type in pandas.

11. You would expect populations to be numeric values. Print out a random sample to try and understand why this is not the case. Do this by running the following commands:

```
np.random.seed(0)
df_populations['population'].sample(10)
```

Here, you have set the **random** seed so that sampling will be reproducible for a given dataset. You should see something similar to the following:

```
110     6,533,500
150     1,902,000
37     38,379,000
75     13,249,924
```

```
109      6,825,442
71       16,244,513
122      5,009,466
73       15,473,818
154      1,454,789
234          3,198
Name: population, dtype: object
```

Here, you can see that the existence of commas in each entry is causing them to be interpreted as strings, instead of integers.

12. Convert the **population** column into a **numeric** datatype by running the following code:

```
df_populations['population'] = df_populations['population']\
                               .str.replace(',','')


df_populations['population'] = \
pd.to_numeric(df_populations['population'], \
            errors='coerce',)
```

First, you use a **string** method to remove any commas and replace them with empty space. Then, you pass over the data again and use the **pd.to_numeric** function to convert each entry into a **numeric** datatype.

13. Did you notice that we set **errors='coerce'** in the preceding code? In order to understand why this was done, pull up the docstring for that function by running **pd.to_numeric?**:

```
Signature: pd.to_numeric(arg, errors='raise', downcast=None)
Docstring:
Convert argument to a numeric type.

The default return dtype is `float64` or `int64`
depending on the data supplied. Use the `downcast` parameter
to obtain other dtypes.

Please note that precision loss may occur if really large numbers
are passed in. Due to the internal limitations of `ndarray`, if
numbers smaller than `-9223372036854775808` (np.iinfo(np.int64).min)
or larger than `18446744073709551615` (np.iinfo(np.uint64).max) are
passed in, it is very likely they will be converted to float so that
they can stored in an `ndarray`. These warnings apply similarly to
`Series` since it internally leverages `ndarray`.
```

Figure 6.22: The docstring for pd.to_numeric

As can be seen, setting **errors='coerce'** will cause any parsing errors to yield **NaNs**. This is in contrast to the default behavior of **pd.to_numeric**, which raises errors when parsing fails for any data entry.

14. Moving on, the next column to deal with is population percent, which should also be a numeric datatype. Print out a random sample of that column by running the following commands:

```
np.random.seed(0)
df_populations['population_pct'].sample(10)
```

You should see something similar to the following:

```
110        0.0838%
150        0.0244%
37          0.492%
75          0.170%
109        0.0875%
71          0.208%
122        0.0642%
73          0.198%
154        0.0187%
234     0.0000410%
Name: population_pct, dtype: object
```

Here, the existence of percent signs (**%**) in each entry is causing them to be interpreted as strings, instead of integers.

15. Convert the population percent column into a **numeric** datatype by running the following code:

```
df_populations['population_pct'] = df_populations['population_pct']\
                    .str.replace('%','')

df_populations['population_pct'] = \
pd.to_numeric(df_populations['population_pct'], \
            errors='coerce',)
```

16. The final column to fix is the **date** column, which was also loaded as a string. It's very normal for pandas to load dates as strings unless instructed otherwise.

Similar to the preceding commands, you can convert the date entries from **string** into **datetime** objects using **pd.to_datetime**. Do this by running the following code:

```
df_populations['date'] = pd.to_datetime(df_populations['date'], \
                                         errors='coerce',)
```

17. Print the **datatypes** of the table by running **df_populations.dtypes**. This should output the following:

```
country                  object
population                int64
population_pct           float64
date             datetime64[ns]
dtype: object
```

18. Look for missing records by running the following command:

```
df_populations.isnull().sum()
```

The output should be similar to the following:

```
country          4
population       1
population_pct   1
date             0
dtype: int64
```

19. Now, deal with this missing data. Identify the rows that have missing data and print them out by running the following code:

```
missing_mask = df_populations.isnull().any(axis=1)
df_populations[missing_mask]
```

This should display the missing data as follows:

	country	population	population_pct	date
0	NaN	NaN	NaN	NaT
123	NaN	4.976684e+06	0.063800	2019-07-01
219	NaN	3.533400e+04	0.000458	2017-01-01
242	NaN	7.800384e+09	100.000000	2020-07-20

Figure 6.23: Output showing the missing data

20. You are simply going to drop these missing rows. In order to do this, first select the indices to drop. Looking back at the row indices displayed in the preceding table, these appear to be rows **121**, **190**, and **218**. Select these dynamically by running the following code:

```
drop_indices = df_populations.index[missing_mask]
drop_indices
```

This should output the expected list of indices:

```
Int64Index([0, 123, 219, 242], dtype='int64')
```

21. Drop these rows by running the code:

```
df_populations = df_populations.drop(drop_indices)
```

Now that it's been cleaned up, you can write our resulting table to file by running the following command:

```
f_name = '../data/countries/populations.csv'
df_populations.to_csv(f_name, index=False)
```

22. Having processed the populations table, now do the same for the interest rates table. Start out by making sure the max row limit on pandas is large enough to view the entire table. Then, print it out by running **df_int_rates**:

	Country or currency union	Central bank interest rate (%)	Date of last change	Average inflation rate (%)
0	Albania	1.00	6 June 2016	1.75
1	Angola	15.50	24 May 2019	17.54
2	Argentina	38.00	5 March 2020	31.17
3	Armenia	5.75	29 January 2019	2.38
4	Australia	0.25	19 March 2020	1.93
5	Azerbaijan	7.25	1 May 2020	6.51
6	Bahamas	4.00	22 December 2016	0.97
7	Bahrain	2.50	31 July 2019	2.40
8	Bangladesh	6.00	6 April 2018	6.38
9	Barbados	7.00	1 June 2009	1.67

Figure 6.24: Displaying the full interest rates table in Jupyter

Look at the column names by running **df_int_rates.columns**.

You should see the following output:

```
Index(['Country or currency union', 'Central bank interest rate (%)',
       'Date of last change', 'Average inflation rate (%)'],
      dtype='object')
```

23. Now, set these manually by running the following code:

```
df_int_rates.columns = ['country', 'interest_rate_pct', \
                        'date_of_last_change', \
                        'average_inflation_rate_pct']
```

24. Print the datatypes of each column by running **df_int_rates.dtypes**.

You should see the following output:

```
country                       object
interest_rate_pct             object
date_of_last_change           object
average_inflation_rate_pct    float64
dtype: object
```

25. The interest rate column should be a **numeric** datatype. Print a sample of the entries to see what might be the problem here. Do this by running the following code:

```
np.random.seed(0)
df_int_rates['interest_rate_pct'].unique()
```

Here's the output showing sample entries:

```
array(['1.00', '15.50', '68.00', '5.75', '0.75', '8.25', '4.00',
       '2.50', '6.00', '7.00', '9.50', '5.00', '4.25', '0.00',
       '1.75', '5.50', '2.95', '4.20', '2.00', '9.00', '3.00',
       '-0.75', '5.25', '6.50', '13.25', '0.50', '20.00', '0.90',
       '3.25', '5.40', '18.00', '0.25', '-0.10', '3.75', '10.00',
       '2.75', '13.50', '7.50', '8.00', '11.00', '2.25', '12.75',
       '1.50', '0.19', '13.00', '1.25', '7.25', '-0.25', '1.375',
       '9.25', '7.75', '11.25', '-', '16.00', '10.25'],
      dtype=object)
```

26. Here, you can see one particular bad value that's made its way into the column: the – sign. Otherwise, the data looks pretty good. Use **pd.to_numeric** to convert this column:

```
df_int_rates['interest_rate_pct'] = \
pd.to_numeric(df_int_rates['interest_rate_pct'], \
              errors='coerce',)
```

The bad value you just identified, **–**, will be converted into a **NaN** entry because we have set **errors=because**.

27. Convert the date entries into **datetime** objects by running the following code:

```
df_int_rates['date_of_last_change'] = \
pd.to_datetime(df_int_rates['date_of_last_change'])
```

28. Print the datatypes of our cleaned up table by running **df_int_rates.dtypes**. This should output the following:

```
country                       object
interest_rate_pct            float64
date_of_last_change    datetime64[ns]
average_inflation_rate_pct   float64
dtype: object
```

29. Look for missing records and deal with them, just like you did for the populations table. Do this by running the following command:

```
df_int_rates.isnull().sum()
```

You should see output similar to the following:

```
country                       1
interest_rate_pct             1
date_of_last_change           2
average_inflation_rate_pct    8
dtype: int64
```

30. Again, like you did for the populations table, select any rows that have missing entries and look at them. Do this by running the following code:

```
missing_mask = df_int_rates.isnull().any(axis=1)
df_int_rates[missing_mask]
```

The following screenshot shows the table with the missing data:

	country	interest_rate_pct	date_of_last_change	average_inflation_rate_pct
16	Central African States	2.950	2017-03-22	NaN
20	Colombia	4.250	2018-04-27	NaN
22	Croatia	3.000	2017-09-17	NaN
25	Eastern Caribbean	6.500	NaT	NaN
27	NaN	0.000	2016-03-10	NaN
35	India	3.750	NaT	6.12
77	Taiwan	1.375	2016-06-30	NaN
88	Uruguay	NaN	2013-06-27	8.39
89	Uzbekistan	16.000	2018-09-22	NaN
91	West African States	2.500	2013-09-16	NaN

Figure 6.25: Displaying the rows with missing values

31. Since many of these rows have good information intact, you would generally want to keep them around. You may also want to go back to earlier work and adjust how the data was scraped so that you are able to capture this missing data.

    For the purposes of this book, however, you are going to drop this data. That way, you will have a very clean dataset to work with in the final activity. Select the indices of the rows to drop by running the following code:

    ```
    drop_indices = df_int_rates.index[missing_mask]
    drop_indices
    ```

    This should output the row numbers as expected, based on the preceding table:

    ```
    Int64Index([16, 20, 22, 25, 27, 35, 77, 88, 89, 91], dtype='int64')
    ```

32. Drop these rows by running the following code:

    ```
    df_int_rates = df_int_rates.drop(drop_indices)
    ```

33. Now that it's been cleaned up, you can write your resulting table to file by running the cell containing the following code:

```
f_name = '../data/countries/ interest_rates.csv'
df_populations.to_csv(f_name, index=False)
```

> **NOTE**
>
> To access the source code for this specific section, please refer to
> https://packt.live/2ACHg63.
>
> You can also run this example online at https://packt.live/2zDrqYu.

This concludes the exercise on processing data for analysis with pandas. We loaded the raw data in the format it was extracted in from Wikipedia, and then processed the tables in order to clean up the data fields prior to analysis. This involved altering column names, checking for null values, converting datatypes, and dropping missing data.

In the next exercise, we'll merge the two tables we processed here.

EXERCISE 6.05: MERGING DATA WITH PANDAS

We're just about ready to start exploring the data that was cleaned in the previous exercise. The final step we'll need to do is merge the two tables. This can be done by joining the rows from each table on the country. We'll load the processed data from previously saved files and merge them into a single DataFrame that will then be used as the data source for our final analysis activity. Follow these steps to complete this exercise:

1. Start up one of the following platforms for running Jupyter Notebooks:

   JupyterLab (run **jupyter lab**)

   Jupyter Notebook (run **jupyter notebook**)

   Then, open up the platform you chose in your web browser by copying and pasting the URL, as prompted in the Terminal.

2. Run the following code to load some of the libraries that you will use to configure our plot settings for the Notebook:

```
import pandas as pd
import numpy as np
import datetime
import time
import os

import matplotlib.pyplot as plt
%matplotlib inline
import seaborn as sns

%config InlineBackend.figure_format='retina'
sns.set() # Revert to matplotlib defaults
plt.rcParams['figure.figsize'] = (9, 6)
plt.rcParams['axes.labelpad'] = 10
sns.set_style("darkgrid")

%load_ext watermark
%watermark -d -v -m -p numpy,pandas,matplotlib,seaborn
```

3. Reload the processed datasets by running the following code:

```
df_int_rates = pd.read_csv('../data/countries/interest_rates.csv')
df_populations = pd.read_csv('../data/countries/populations.csv')
```

4. Run the following command to print the **columns** table of **df_int_rates**:

```
df_int_rates.columns
```

The output is as follows:

```
Index(['country', 'interest_rate_pct', 'date_of_last_change',
       'average_inflation_rate_pct'],
      dtype='object')
```

Run the following command to print the **columns** table of **df_populations**:

```
df_populations.columns
```

The output is as follows:

```
Index(['country', 'population', 'population_pct',
        'date'], dtype='object')
```

5. Looking at the columns for each table, you can see that they should be joined on the country key. Perform an outer merge by running the following code:

```
df_merge = pd.merge(df_populations, df_int_rates, \
                    left_on='country', right_on='country', \
                    how='outer')
df_merge
```

This will output a display of the merged table:

	country	population	population_pct	date	interest_rate_pct	date_of_last_change	average_inflation_rate_pct
0	China	1.403613e+09	1.800000e+01	2020-07-20	4.20	2019-09-20	1.93
1	India	1.364925e+09	1.750000e+01	2020-07-20	NaN	NaN	NaN
2	United States	3.299857e+08	4.230000e+00	2020-07-20	0.25	2020-05-06	2.33
3	Indonesia	2.696034e+08	3.460000e+00	2020-07-01	4.50	2020-03-19	5.29
4	Pakistan	2.208923e+08	2.830000e+00	2020-07-01	9.00	2020-04-17	5.03
5	Brazil	2.118177e+08	2.720000e+00	2020-07-20	2.25	2020-06-17	6.73

Figure 6.26: The merged populations and interest rates table

6. When doing this **merge**, you lost some important context on the **date** columns. Fix this by renaming those columns, running the following code:

```
column_map = {'date': 'date_population_update', \
              'date_of_last_change': \
              'date_interest_rate_last_change'}

df_merge = df_merge.rename(columns=column_map)
```

Here, you have defined how the columns should be mapped and then executed the logic by using the **rename** function on our DataFrame.

7. Run the **df_merge.head()** command to print the head of our DataFrame and confirm the column names have been altered, as expected:

	country	population	population_pct	date_population_update	interest_rate_pct	date_interest_rate_last_change	average_inflation_rate_pct
0	China	1.403613e+09	18.00	2020-07-20	4.20	2019-09-20	1.93
1	India	1.364925e+09	17.50	2020-07-20	NaN	NaN	NaN
2	United States	3.299857e+08	4.23	2020-07-20	0.25	2020-05-06	2.33
3	Indonesia	2.696034e+08	3.46	2020-07-01	4.50	2020-03-19	5.29
4	Pakistan	2.208923e+08	2.83	2020-07-01	9.00	2020-04-17	5.03

Figure 6.27: The merged table head, showing the updated column names

Since you performed an outer merge, pandas will use all the data from both tables and insert NaN entries for keys that don't align between the two.

8. Print the number of NaN entries in the merged table by running the following command:

```
df_merge.isnull().sum()
```

This should output the following:

```
country                             0
population                          1
population_pct                      1
date_population_update              1
interest_rate_pct                 157
date_interest_rate_last_change    157
average_inflation_rate_pct        157
dtype: int64
```

9. Now, drop these rows. First, run **len(df_merge)** to print the current length of the table. Then, print out a few sample records by running the following command:

```
df_merge[df_merge.isnull().any(axis=1)].sample(10)
```

Current length of the table is **240**. Here's the output showing our merged tables:

	country	population	population_pct	date_population_update	interest_rate_pct	date_interest_rate_last_change	average_inflation_rate_pct
27	Colombia	50372424.0	0.646000	2020-06-30	NaN	NaN	NaN
91	United Arab Emirates	9890400.0	0.127000	2020-07-01	NaN	NaN	NaN
195	U.S. Virgin Islands	104578.0	0.001340	2019-07-01	NaN	NaN	NaN
202	Dominica	71808.0	0.000921	2019-07-01	NaN	NaN	NaN
124	Liberia	4568298.0	0.058600	2020-07-01	NaN	NaN	NaN
204	Bermuda	64027.0	0.000821	2019-07-01	NaN	NaN	NaN
203	Cayman Islands	65813.0	0.000844	2018-12-31	NaN	NaN	NaN
221	Palau	17900.0	0.000229	2018-07-01	NaN	NaN	NaN
191	Aruba	112190.0	0.001440	2019-12-31	NaN	NaN	NaN
169	Suriname	587000.0	0.007530	2020-07-01	NaN	NaN	NaN

Figure 6.28: A sample of rows with missing values

10. You can see that many of the rows that contain population data do not have interest rate data. In order to drop rows that have any missing values, run the following command:

```
df_merge = df_merge.dropna()
```

11. Then, confirm that all the missing entries have been dropped by running the following command:

```
df_merge.isnull().sum()
```

This should output the following:

```
country                         0
population                      0
population_pct                  0
date_population_update          0
interest_rate_pct               0
date_interest_rate_last_change  0
average_inflation_rate_pct      0
dtype: int64
```

12. Print the new length by running **len(df_merge)**. Note how it's much shorter than before.

13. Lastly, having cleaned up our dataset in preparation for analysis, write the table to a CSV file by running the following code:

```
f_name = '../data/countries/country_data_merged.csv'
df_merge.to_csv(f_name, index=False)
```

> **NOTE**
>
> To access the source code for this specific section, please refer to
> https://packt.live/2ACHg63.
>
> You can also run this example online at https://packt.live/2zDrqYu.

So far in this chapter's exercises, we have gathered country population and interest rate datasets from Wikipedia and cleaned them up using pandas and Jupyter Notebooks. This was all done in a reproducible environment that could serve as the backbone of a data pipeline. A positive implication of this is being able to return to your analysis after having forgotten about the details, easily understand how it's working, and debug new issues that may arise as you step through the notebook. In addition, you could share this work with colleagues who can reproduce your results from scratch. This notebook could also serve as the foundation for a data pipeline that you may wish to run periodically to extract and process the most current data.

ACTIVITY 6.02: ANALYZING COUNTRY POPULATIONS AND INTEREST RATES

The aim of this activity is to give you an opportunity to apply the analysis techniques you've learned about in this book, such as data loading, visualization, and exploratory analysis.

You'll start this activity by loading the processed data that has been gathered and processed previously. By the end of this activity, you will have a deeper understanding of some of the key information in the dataset, and an array of different visualizations of the data. Follow these steps to complete this exercise:

1. Create a new Jupyter Notebook and load the necessary libraries.

2. Load the table that we processed in the previous exercise into a pandas DataFrame with a variable name of **df**. Recall that this data was saved to the **country_data_merged.csv** file. Print the head of the table to check that it looks as you expect it to.

3. Print the datatypes of each column. Do you notice any issues?

4. Convert the **date** column datatypes into **datetimes**.

5. Check if there's any missing data in the table.

6. Plot and compare the histograms for the **date** columns, that is, **date_population_update** and **date_interest_rate_last_change**.

7. What country in the dataset has the oldest date since the last interest rate change?

8. Do any countries in the dataset have negative interest rates? If so, then which ones?

9. Plot a bar chart of the top countries by population.

10. Plot a scatter chart of **population_pct** versus **interest_rate_pct**.

11. Can you see any outliers along the population axis? If so, then what are they?

12. Plot a scatter chart of **average_inflation_rate_pct** versus **interest_rate_pct**.

> **NOTE**
>
> The solution to this activity can be found on page 309. The solution also includes some bonus material that shows how you can take the analysis further by fitting a clustering model on the data. Feel free to explore the bonus material and use it as a starting point for your own further research.

SUMMARY

In this chapter, we worked through the process of pulling tables from *Wikipedia* using web scraping techniques, cleaning up the resulting data with pandas, and producing a final analysis.

We started by looking at how **HTTP** requests work, focusing on **GET** requests and their response status codes. Then, we went into the Jupyter Notebook and made HTTP requests with Python using the **requests** library. We saw how Jupyter can be used to render HTML in the notebook, along with actual web pages that can be interacted with. In order to learn about web scraping, we saw how **BeautifulSoup** can be used to parse text from the HTML, and used this library to scrape tabular data from Wikipedia.

After pulling two tables of data, we processed them for analysis with pandas. The first table contained the central bank interest rates for each country, while the second table contained the populations. We combined these into a single table that was then used for the final analysis, which involved creating visualizations to find patterns.

Now, let's take a step back from this chapter specifically and review what we have learned since the start of this book.

In the first chapter, we walked through an overview of the data science landscape in order to set the context for the remainder of this book. We learned about the origin of data science and introduced key concepts in the field, such as being able to plan out data-driven solutions to business problems and understanding the roles of data visualization, exploration, and modeling. We also looked at various career paths, ways to get involved in the community, and discussed some challenges presented by data science.

Next, we introduced Jupyter Notebooks and learned why they are useful for data analysis and modeling. We ran through two interfaces for running them: the older Jupyter Notebook platform and the newer Jupyter Lab. Then, we ran through a sample notebook to explore how they work and learn about Jupyter features such as magic functions and tab completion.

Having learned the basic functionality of Jupyter, we started working through a notebook where we studied the `Boston Housing` dataset. This was our first time using Jupyter to study real-world style problems, where we explored the dataset using the pandas library and created a handful of visualizations and simple models.

After we were finished with the `Boston Housing` dataset, we turned our attention to a key component of data science: predictive modeling. We expanded on some of the ideas we presented in the opening chapter by discussing the steps for planning a modeling strategy in detail. We also learned about another key step in modeling data, which is the data preparation stage, which is necessary before using training algorithms. Among other considerations, this included filling missing data, converting from categorical into numeric features, and splitting data into training and testing sets.

At this point, we introduced the Human Resource Analytics dataset for the *Employee Retention Problem*. After identifying a modeling strategy around this dataset, we applied data processing techniques to clean it up so that it was ready to be used to train machine learning models. We started by focusing on just two features in the dataset and learned about training simple models such as **Support Vector Machines (SVMs)** and **k-Nearest Neighbors (KNN)** classifiers. In particular, we paid close attention to the decision boundaries that were formed by such models and the effects of overfitting.

Having worked through the basics of modeling, we then introduced some more advanced concepts in machine learning, including k-fold cross-validation and validation curves. Applying these to our problem, we attempted to increase our model accuracy by optimizing hyperparameters such as the max depth of our Random Forest model.

Finally, we learned about the dimensionality reduction technique known as PCA and applied it to our problem in order to help boost the overall accuracy of our model.

Finally, in the last part of this book, we shifted briefly from data modeling and analysis to learn about data collection from the web. In particular, we learned about making web requests with Python in order to call Web APIs or HTML resources and parse data from the response. We finished up this section by applying what we learned earlier in this book to explore, visualize, and model the data we collected.

Whether or not you were experienced with Python or Jupyter coming into this book, hopefully you've learned plenty of useful and applicable skills for approaching and solving data science problems.

As you move forward on your learning path, you're encouraged to continue building on the topics that piqued your interest in this book. You may wish to seek out resources such as those discussed in the article, *Data Science at a Glance*, and work through data science projects of your own from start to finish.

As a student of data science or a professional, you will continuously be presented with difficult problems that must be overcome. Persevering in the face of adversity is not something that can be taught, but instead must come from within. So, if you ever feel like giving up, perhaps you can be motivated by the thought that few things worthwhile are ever easy.

APPENDIX

CHAPTER 1: INTRODUCTION TO JUPYTER NOTEBOOKS

ACTIVITY 1.01: USING JUPYTER TO LEARN ABOUT PANDAS DATAFRAMES

Solution:

1. Start one of the following platforms to run Jupyter Notebooks:

    Jupyter Notebook (run **jupyter notebook**)

    JupyterLab (run **jupyter lab**)

    Then, open the platform in your web browser by copying and pasting the URL, as prompted in the Terminal.

2. Load the **numpy** library as follows:

```
import numpy as np
```

3. Import **pandas**, as follows:

```
import pandas as pd
```

4. Pull up the docstring for the pandas DataFrame object, as follows:

```
pd.DataFrame?
```

    The output is as follows:

```
Init signature: pd.DataFrame(data=None, index=None, columns=None, dtype=None, copy=False)
Docstring:
Two-dimensional size-mutable, potentially heterogeneous tabular data
structure with labeled axes (rows and columns). Arithmetic operations
align on both row and column labels. Can be thought of as a dict-like
container for Series objects. The primary pandas data structure.

Parameters
----------
data : ndarray (structured or homogeneous), Iterable, dict, or DataFrame
    Dict can contain Series, arrays, constants, or list-like objects

    .. versionchanged :: 0.23.0
       If data is a dict, column order follows insertion-order for
       Python 3.6 and later.

    .. versionchanged :: 0.25.0
       If data is a list of dicts, column order follows insertion-order
       for Python 3.6 and later.

index : Index or array-like
    Index to use for resulting frame. Will default to RangeIndex if
    no indexing information part of input data and no index provided
```

Figure 1:39: The docstring for pd.DataFrame

5. Use a dictionary to create a DataFrame with **fruit** and **score** columns, as follows:

```
fruit_scores = {'fruit': ['apple', 'orange', \
                          'banana', 'blueberry'], \
                'score': [4, 2, 9, 8],}
df = pd.DataFrame(data=fruit_scores)
```

The DataFrame is as follows:

	fruit	score
0	apple	4
1	orange	2
2	banana	9
3	blueberry	8

Figure 1.40: A DataFrame with fruits and their scores

6. Use tab completion to pull up a list of functions available for the DataFrame by typing **df.** and pressing *Tab*. The list of functions should then appear as a list of autocomplete options, as follows:

```
# Use tab completion to pull up a list of functions for df
# df.<tab>

df.|
```

f	abs	function
f	add	function
f	add_prefix	function
f	add_suffix	function
f	agg	function
f	aggregate	function
f	align	function
f	all	function
f	any	function
f	append	function

Figure 1.41: Example of the tab help feature in Jupyter

7. Pull up the docstring for **sort_values** by running a cell with the following code:

```
df.sort_values?
```

The output is as follows:

```
Signature:
df.sort_values(
    by,
    axis=0,
    ascending=True,
    inplace=False,
    kind='quicksort',
    na_position='last',
)
Docstring:
Sort by the values along either axis.

Parameters
----------
        by : str or list of str
            Name or list of names to sort by.

            - if `axis` is 0 or `'index'` then `by` may contain index
              levels and/or column labels
            - if `axis` is 1 or `'columns'` then `by` may contain column
              levels and/or index labels
```

Figure 1.42: The docstring for pd.DataFrame.sort_values

8. Sort the DataFrame by score in descending order, as follows:

```
df.sort_values(by='score', ascending=False)
```

The output is as follows:

	fruit	score
2	banana	9
3	blueberry	8
0	apple	4
1	orange	2

Figure 1.43: Sorted fruits DataFrame

9. This will compute and display the sorted DataFrame; however, the original sorting on **df** will remain intact.

10. See how long it takes to compute the preceding sorting operation, as follows:

```
%timeit df.sort_values(by='score', ascending=False)
```

The output is as follows:

```
349 µs ± 6.43 µs per loop (mean ± std. dev. of 7 runs, 1000 loops
each)
```

> **NOTE**
>
> To access the source code for this specific section, please refer to https://packt.live/3ftGze0.
>
> You can also run this example online at https://packt.live/2Y49zTQ.

CHAPTER 2: DATA EXPLORATION WITH JUPYTER

ACTIVITY 2.01: BUILDING A THIRD-ORDER POLYNOMIAL MODEL

Solution:

1. Load the necessary libraries and the dataset from scikit-learn, as follows:

```
import pandas as pd
import matplotlib.pyplot as plt
import numpy as np
from sklearn import datasets
boston = datasets.load_boston()
df = pd.DataFrame(data=boston['data'], \
                  columns=boston['feature_names'],)
df['MEDV'] = boston['target']
```

2. First, we will pull out our dependent feature and target variable from **df**, as follows:

```
y = df['MEDV'].values
x = df['LSTAT'].values.reshape(-1,1)
```

This is identical to what we did earlier for the linear model.

3. Verify what **x** looks like by executing the following code:

```
x[:3]
```

The output is as follows:

```
array([[4.98],
       [9.14],
       [4.03]])
```

Notice how each element in the array is itself an array with length 1. This is what **reshape(-1,1)** does, and it is the form expected by scikit-learn.

4. Transform **x** into polynomial features by importing the appropriate transformation tool from scikit-learn and instantiating the third-degree polynomial feature transformer:

```
from sklearn.preprocessing import PolynomialFeatures
poly = PolynomialFeatures(degree=3)
```

The rationale for this step may not be immediately obvious but will be explained shortly.

The representation of **poly** is as follows:

```
PolynomialFeatures(degree=3, include_bias=True,
                   interaction_only=False, order='C')
```

5. Transform the **LSTAT** feature (as stored in the **x** variable) by running the **fit_ transform** method, as follows:

```
x_poly = poly.fit_transform(x)
```

Here, we have used the instance of the transformer feature to transform the **LSTAT** variable.

6. Verify what **x_poly** looks like by using the following code:

```
x_poly[:3]
```

The output is as follows:

```
array([[  1.     ,    4.98  ,    24.8004 ,  123.505992],
       [  1.     ,    9.14  ,    83.5396 ,  763.551944],
       [  1.     ,    4.03  ,    16.2409 ,   65.450827]])
```

Unlike x, the arrays in each row now have a length of 4, where the values have been calculated as x^0, x^1, x^2, and x^3.

Now, we are going to use this data to fit a linear model. Labeling the features as a, b, c, and d, we will calculate the coefficients a_0, a_1, a_2, and a_3 of the linear model:

$$y = a_0 a + a_1 b + a_2 c + a_3 d$$

Figure 2.28: Linear model equation

We can plug in the definitions of a, b, c, and d to get the following polynomial model, where the coefficients are the same as the previous ones:

$$y = a_0 + a_1 x + a_2 x^2 + a_3 x^3$$

Figure 2.29: Third-order polynomial equation

7. Import the **LinearRegression** class and train a linear classification mode by running the following command:

```
from sklearn.linear_model import LinearRegression
clf = LinearRegression(fit_intercept=False)
clf.fit(x_poly, y)
```

The output is as follows:

```
LinearRegression(copy_X=True, fit_intercept=False,
                 n_jobs=None, normalize=False)
```

8. Extract the coefficients and print the polynomial model using the following code:

```
x_0, x_1, x_2, x_3 = clf.coef_
msg = ('model: y = {:.3f} + {:.3f}x + {:.3f}x^2 + {:.3f}x^3'\
       .format(x_0, x_1, x_2, x_3))
print(msg)
```

The output is as follows:

```
model: y = 48.650 + -3.866x + 0.149x^2 + -0.002x^3
```

Determine the predicted values for each sample and calculate the residuals by running the following code:

```
y_pred = clf.predict(x_poly)
resid_MEDV = y - y_pred
```

9. Print the first 10 residual values, as follows:

```
resid_MEDV[:10]
```

The output is as follows:

```
array([-8.84025736, -2.61360313, -0.65577837, -5.11949581,
        4.23191217, -3.56387056,  3.16728909, 12.00336372,
        4.03348935,  2.87915437])
```

We'll plot these soon in order to compare them with the linear model residuals, but first, we will calculate the MSE.

10. Run the following code to print the MSE for the third-order polynomial model:

```
from sklearn.metrics import mean_squared_error
error = mean_squared_error(y, y_pred)
```

11. Print the MSE, as follows:

```
print('mse = {:.2f}'.format(error))
```

The output is as follows:

```
mse = 28.88
```

As can be seen, the **MSE** is significantly less for the polynomial model compared to the linear model (which was 38.5). This error metric can be converted into an average error in dollars by taking the square root. Doing this for the polynomial model, we find that the average error for the median house value is only $5,300.

Now, we'll visualize the model by plotting the polynomial line of best fit along with the data.

12. Plot the polynomial model along with the samples, as follows:

```
fig, ax = plt.subplots()
# Plot the samples
ax.scatter(x.flatten(), y, alpha=0.6)

# Plot the polynomial model
x_ = np.linspace(2, 38, 50).reshape(-1, 1)
x_poly = poly.fit_transform(x_)
y_ = clf.predict(x_poly)
ax.plot(x_, y_, color='red', alpha=0.8)

ax.set_xlabel('LSTAT')
ax.set_ylabel('MEDV')
plt.savefig('../figures/chapter-2-boston-housing-poly.png', \
            bbox_inches='tight', dpi=300,)
```

The output is as follows:

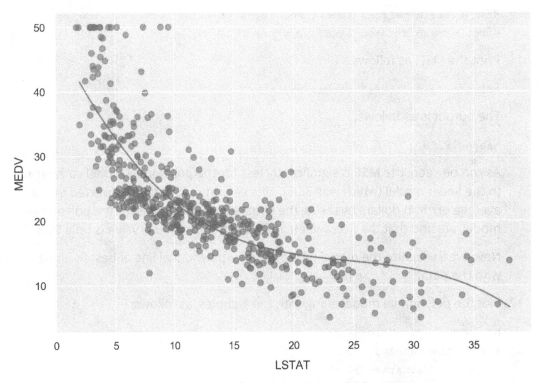

Figure 2.30: Plot of the polynomial model for MEDV

Here, we are plotting the red curve by calculating the polynomial model predictions on an array of **x** values. The array of **x** values was created using `np.linspace`, resulting in 50 values arranged evenly between 2 and 38.

Now, we'll plot the corresponding residuals. Although we used seaborn for this earlier, we'll have to do this manually to show results for a scikit-learn model. Since we already calculated the residuals, as referenced by the **resid_MEDV** variable, we simply need to plot this list of values on a scatter chart.

13. Plot the residuals by running the following code:

```
fig, ax = plt.subplots(figsize=(5, 7))
ax.scatter(x, resid_MEDV, alpha=0.6)
ax.set_xlabel('LSTAT')
ax.set_ylabel('MEDV Residual $(y-\hat{y})$')
plt.axhline(0, color='black', ls='dotted')
plt.savefig('../figures/chapter-2-boston-housing-'\
            'poly-residuals.png', \
            bbox_inches='tight', dpi=300,)
```

This will result in the following output:

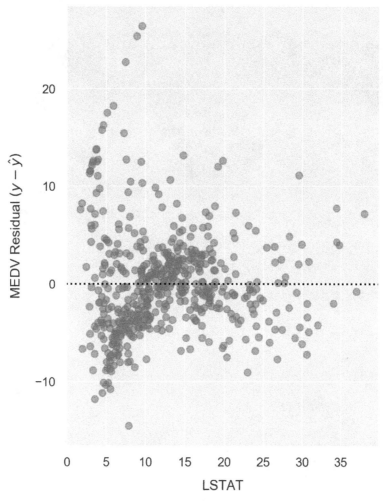

Figure 2.31: Plot of the polynomial model residuals

Compared to the linear model's **LSTAT** residual plot, the polynomial model residuals appear to be more closely clustered around $y - \hat{y} = 0$. Note that y is the sample **MEDV** and that $\hat{y}$ is the predicted value. There are still clear patterns, such as the cluster near $x = 7$ and $y = -7$ that indicates suboptimal modeling.

> **NOTE**
>
> To access the source code for this specific section, please refer to https://packt.live/2UIzwq8.
>
> You can also run this example online at https://packt.live/37DzuVK.

CHAPTER 3: PREPARING DATA FOR PREDICTIVE MODELING

ACTIVITY 3.01: PREPARING TO TRAIN A PREDICTIVE MODEL FOR EMPLOYEE RETENTION

Solution:

1. Check the head of the table by running the following command:

```
%%bash
head ../data/hr-analytics/hr_data.csv
```

Note how we specify paths relative to the notebook's location. In this case, we need to step back one directory (by using " . . " in the file path), which brings us to the root folder for the project. Then, we look in **data/hr-analytics** for **hr_data.csv**.

This will generate the following output:

```
satisfaction_level,last_evaluation,number_project,average_montly_hours,time_spend_company,work_accid
ent,left,promotion_last_5years,is_smoker,department,salary
0.38,0.53,2,157,3,0,yes,0,,sales,low
0.8,0.86,5,262,6,0,yes,0,yes,sales,medium
0.11,0.88,7,272,4,0,yes,0,,sales,medium
0.72,0.87,5,223,5,0,yes,0,,sales,low
0.37,0.52,2,,,0,yes,0,no,sales,low
0.41,0.5,2,,,0,yes,0,,sales,low
0.1,0.77,6,247,4,0,yes,0,,sales,low
0.92,0.85,5,259,5,0,yes,0,,sales,low
0.89,1,5,224,5,0,yes,0,,sales,low
```

Figure 3.25: Printing the head of hr_data.csv with bash

2. If you cannot run bash in your notebook, run the following command:

```
with open('../data/hr-analytics/hr_data.csv', 'r') as f:
    for _ in range(10):
        print(next(f).strip())
```

The output is as follows:

```
satisfaction_level,last_evaluation,number_project,average_montly_hours,time_spend_company,work_accid
ent,left,promotion_last_5years,is_smoker,department,salary
0.38,0.53,2,157,3,0,yes,0,,sales,low
0.8,0.86,5,262,6,0,yes,0,yes,sales,medium
0.11,0.88,7,272,4,0,yes,0,,sales,medium
0.72,0.87,5,223,5,0,yes,0,,sales,low
0.37,0.52,2,,,0,yes,0,no,sales,low
0.41,0.5,2,,,0,yes,0,,sales,low
0.1,0.77,6,247,4,0,yes,0,,sales,low
0.92,0.85,5,259,5,0,yes,0,,sales,low
0.89,1,5,224,5,0,yes,0,,sales,low
```

Figure 3.26: Printing the head of hr_data.csv with Python

Judging by the output, convince yourself that it looks to be in standard CSV format. For CSV files, we should be able to simply load the data with **pd.read_csv**.

3. Load the data with pandas, as follows:

```
df = pd.read_csv('../data/hr-analytics/hr_data.csv')
```

You should write this out yourself and try using tab completion to help type the file path.

4. Inspect the columns as follows:

```
df.columns
```

The output is as follows:

```
Index(['satisfaction_level', 'last_evaluation', 'number_project',
       'average_montly_hours', 'time_spend_company', 'work_accident', 'left',
       'promotion_last_5years', 'is_smoker', 'department', 'salary'],
      dtype='object')
```

Figure 3.27: Columns loaded from hr_data.csv

5. Ensure that the data has loaded as expected by printing the head of the DataFrame, as follows:

```
df.head()
```

The output is similar to the following:

	satisfaction_level	last_evaluation	number_project	average_montly_hours	time_spend_company	work_accident	left
0	0.38	0.53	2	157.0	3.0	0	yes
1	0.80	0.86	5	262.0	6.0	0	yes
2	0.11	0.88	7	272.0	4.0	0	yes
3	0.72	0.87	5	223.0	5.0	0	yes
4	0.37	0.52	2	NaN	NaN	0	yes

Figure 3.28: The head of hr_data.csv in a pandas DataFrame

> **NOTE**
>
> The preceding DataFrames are cropped for representation purpose. The complete output can be found here: https://packt.live/2YEsiUX.

6. Print the tail of the DataFrame, as follows:

```
df.tail()
```

The output is as follows:

	satisfaction_level	last_evaluation	number_project	average_montly_hours	time_spend_company	work_accident	left
14994	0.40	0.57	2	151.0	3.0	0	yes
14995	0.37	0.48	2	160.0	3.0	0	yes
14996	0.37	0.53	2	143.0	3.0	0	yes
14997	0.11	0.96	6	280.0	4.0	0	yes
14998	0.37	0.52	2	158.0	3.0	0	yes

Figure 3.29: The tail of hr_data.csv in a pandas DataFrame

We can see that it appears to have loaded correctly. Based on the tail index values, there are nearly 15,000 rows.

> **NOTE**
>
> The preceding DataFrames are cropped for representation purpose. The complete output can be found here: https://packt.live/2YEsiUX.

7. Check the number of rows (including the header) in the CSV file with the following code:

```
with open('../data/hr-analytics/hr_data.csv') as f:
    num_lines = len([line for line in f.read().splitlines()\
                    if line.strip()])
num_lines
```

This will print the following output:

```
15000
```

8. See how many rows are in the DataFrame:

```
len(df)
```

This will print the following output:

```
14999
```

This number is one less than the number of rows in the file, since it does not include the header (column names). Therefore, we can conclude that all the records have been loaded.

9. Check how the **left** target variable is distributed, as follows:

```
df.left.value_counts()
```

This will print the following output:

```
no      11428
yes      3571
Name: left, dtype: int64
```

10. This can be visualized as follows:

```
df.left.value_counts().plot('barh')
plt.show()
```

The output is as follows:

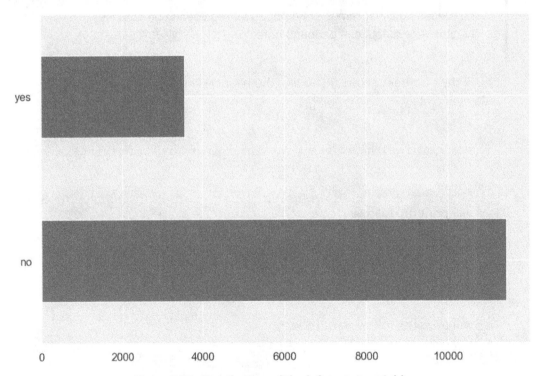

Figure 3.30: Distribution of the left target variable

About three-quarters of the samples are employees who have not left. The group that has left make up the other quarter of the samples. This tells us we are dealing with an imbalanced classification problem, which means we'll have to take special measures to account for each class when calculating accuracies.

11. Check for missing values, as follows:

```
df.left.isnull().sum()
```

This will print the following output:

```
0
```

As can be seen, there are no missing values.

12. Print the data type of each column, as follows:

```
df.dtypes
```

Observe how we have a mix of continuous and discrete features:

```
satisfaction_level        float64
last_evaluation           float64
number_project              int64
average_montly_hours      float64
time_spend_company        float64
work_accident               int64
left                       object
promotion_last_5years       int64
is_smoker                  object
department                 object
salary                     object
dtype: object
```

Figure 3.31: Data types of each column

13. Display the feature distributions by running the following code:

```
for f in df.columns:
    fig = plt.figure()
    s = df[f]
    if s.dtype in ('float', 'int'):
        num_bins = min((30, len(df[f].unique())))
        s.hist(bins=num_bins)
    else:
        s.value_counts().plot.bar()
    plt.xlabel(f)
```

The output is as follows:

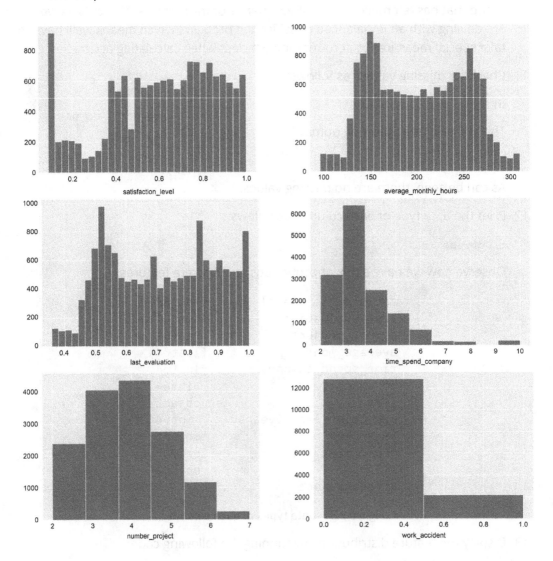

Figure 3.32: Distribution of values for each column (1/2)

The rest of the plots look as follows:

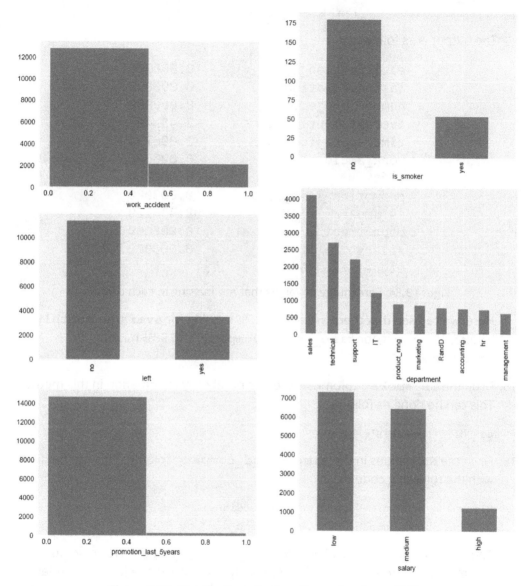

Figure 3.33: Distribution of values for each column (2/2)

For many of the features, we can see a wide distribution over the possible values, indicating a good variety in the feature spaces. This is encouraging; features that are strongly grouped around a small range of values may not be very informative for our model. For example, we can see that this is the case for **promotion_last_5years**, where the vast majority of samples are 0.

14. Check how many **NaN** values are in each column, as follows:

```
df.isnull().sum() / len(df) * 100
```

The output is as follows:

```
satisfaction_level        0.000000
last_evaluation           0.000000
number_project            0.000000
average_montly_hours      2.453497
time_spend_company        1.006734
work_accident             0.000000
left                      0.000000
promotion_last_5years     0.000000
is_smoker                98.433229
department                0.000000
salary                    0.000000
dtype: float64
```

Figure 3.34: Percentage of values that are missing in each column

Here, we can see that there are about 2.5% missing for **average_monthly_hours**, 1% missing for **time_spend_company**, and 98% missing for **is_smoker**.

15. Drop the **is_smoker** column as there is barely any information in this metric. This can be done as follows:

```
del df['is_smoker']
```

16. Fill in the **NaN** values in the **time_spend_company** column. This can be done with the following code:

```
fill_value = df.time_spend_company.median()
df.time_spend_company = df.time_spend_company.fillna(fill_value)
```

The final column to deal with is **average_montly_hours**. We could do something similar to what we did previously and use the median or rounded mean as the integer fill value. Instead, though, let's try to take advantage of its relationship with another variable. This may allow us to fill in the missing data more accurately.

17. Check for the following:

```
df.isnull().sum() / len(df) * 100
```

The output is as follows:

```
satisfaction_level        0.000000
last_evaluation           0.000000
number_project            0.000000
average_montly_hours      2.453497
time_spend_company        0.000000
work_accident             0.000000
left                      0.000000
promotion_last_5years     0.000000
department                0.000000
salary                    0.000000
dtype: float64
```

Figure 3.35: Boxplot showing how the average monthly hours and number of projects are related

18. Make a boxplot of **average_montly_hours** segmented by **number_project**, as follows:

```
sns.boxplot(x='number_project', y='average_montly_hours', data=df)
plt.savefig('../figures/chapter-3-hr-analytics-hours-num-'\
            'proj-boxplot.png', bbox_inches='tight', dpi=300,)
```

> **NOTE**
>
> The path will differ based on where you want to save the image.

The output is as follows:

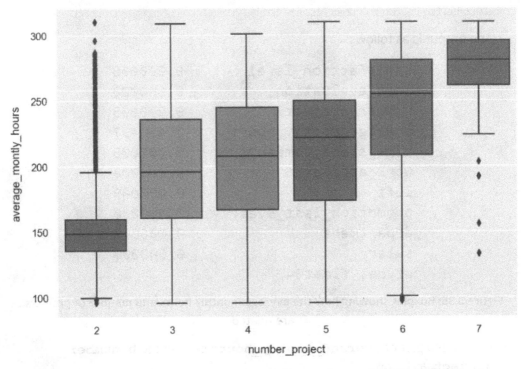

Figure 3.36: Average number of hours worked for each "number of projects" bucket

Here, we can see how the number of projects is correlated with **average_monthly_hours**, a result that is hardly surprising. We'll exploit this relationship by filling in the **NaN** values of **average_montly_hours** based on the number of projects for that record.

Specifically, we'll use the mean of each group.

19. Calculate the mean of each group, as follows:

```
mean_per_project = (df.groupby('number_project') \
                    .average_montly_hours.mean())
mean_per_project = dict(mean_per_project)
mean_per_project
```

The output is as follows:

```
{2: 160.16353543979506,
 3: 197.47882323104236,
 4: 205.07858315740089,
 5: 211.99962839093274,
 6: 238.73947368421054,
 7: 276.015873015873}
```

Then, we can map this onto the **number_project** column and pass the resulting series object as the argument to **fillna**.

20. Fill in the **NaN** values in **average_montly_hours**, as follows:

```
fill_values = df.number_project.map(mean_per_project)
df.average_montly_hours = (df.average_montly_hours\
                           .fillna(fill_values))
```

21. Confirm that **df** has no more **NaN** values by running the following assertion test. If it does not raise an error, then you have successfully removed the **NaN** values from the table:

```
assert df.isnull().sum().sum() == 0
```

22. Transform the string and Boolean fields into integer representations. In particular, we'll manually convert the **left** table variable from **yes** and **no** into **1** and **0** and build the one-hot encoded features. This can be done using the following code:

```
df.left = df.left.map({'no': 0, 'yes': 1})
df = pd.get_dummies(df)
```

23. Show the fields as follows:

```
df.columns
```

The output is as follows:

```
Index(['satisfaction_level', 'last_evaluation', 'number_project',
       'average_montly_hours', 'time_spend_company', 'work_accident', 'left',
       'promotion_last_5years', 'department_IT', 'department_RandD',
       'department_accounting', 'department_hr', 'department_management',
       'department_marketing', 'department_product_mng', 'department_sales',
       'department_support', 'department_technical', 'salary_high',
       'salary_low', 'salary_medium'],
      dtype='object')
```

Figure 3.37: A screenshot of the different fields in the DataFrame

Here, we can see that department and salary have been split into various one-hot encoded features.

The final step to perform in order to prepare our data for machine learning is to scale the features, but it's not appropriate to do this now. We'll learn more about feature scaling when we train our first models on the Human Resource Analytics dataset in the next chapter.

24. We have completed the data preprocessing stage and are ready to move on to training the models. Let's save our preprocessed data by running the following code:

```
df.to_csv('../data/hr-analytics/hr_data_processed.csv', \
          index=False)
```

NOTE

When saving the pandas DataFrame as a CSV file, notice that we pass the `index=False` argument. This is done so that our index is not written to file. In general, the index might contain multiple fields of information, but in our case, it's simply a label range spanning from 0 to the length of our data. Therefore, we should not bother writing this information to the output CSV.

NOTE

To access the source code for this specific section, please refer to https://packt.live/2YEsiUX.

You can also run this example online at https://packt.live/2Y3vvi4.

CHAPTER 4: TRAINING CLASSIFICATION MODELS

ACTIVITY 4.01: TRAINING AND VISUALIZING SVM MODELS WITH SCIKIT-LEARN

SOLUTION:

1. Create a new Jupyter notebook.

2. In the first cell, add the following lines of code to load the libraries we'll be using and set up our plot environment for the notebook:

```
import numpy as np
import datetime
import time
import os

import matplotlib.pyplot as plt
%matplotlib inline
import seaborn as sns

%config InlineBackend.figure_format='retina'
sns.set() # Revert to matplotlib defaults
plt.rcParams['figure.figsize'] = (8, 8)
plt.rcParams['axes.labelpad'] = 10
sns.set_style("darkgrid")
```

3. In the next cell, enter the following code to print the date, version numbers, and hardware information:

```
%load_ext watermark
%watermark -d -v -m -p \
requests,numpy,pandas,matplotlib,seaborn,sklearn
```

You should get the following output:

```
2020-02-13

CPython 3.7.5
IPython 7.10.1

requests 2.22.0
numpy 1.17.4
pandas 0.25.3
matplotlib 3.1.1
seaborn 0.9.0
sklearn 0.21.3

compiler   : Clang 4.0.1 (tags/RELEASE_401/final)
system     : Darwin
release    : 18.7.0
machine    : x86_64
processor  : i386
CPU cores  : 8
interpreter: 64bit
```

Figure 4.19: Output of loading all the required libraries

4. Load the preprocessed Human Resource Analytics dataset by running the following command:

```
df = pd.read_csv('../data/hr-analytics/hr_data_processed.csv')
```

NOTE

The path will vary based on where the data is stored. Provide the absolute path, in case you are working from a different folder.

5. Describe the **number_project** and **average_monthly_hours** features by running the following command:

```
df[['number_project', 'average_montly_hours']].describe()
```

This produces the following output:

	number_project	average_montly_hours
count	14999.000000	14999.000000
mean	3.803054	200.992489
std	1.232592	49.492423
min	2.000000	96.000000
25%	3.000000	156.000000
50%	4.000000	200.000000
75%	5.000000	244.000000
max	7.000000	310.000000

Figure 4.20: Summary description of values for number_project and average_monthly_hours

Comparing the **mean**, **min**, and **max** of each, notice how **number_project** is limited to the range **2 – 7**, whereas **average_monthly_hours** ranges from **96 – 310**.

6. Split the **number_project** and **average_monthly_hours** features into training and testing sets by running the following code:

```
from sklearn.model_selection import train_test_split

features = ['number_project', 'average_montly_hours']
X_train, X_test, \
y_train, y_test = train_test_split(df[features].values, \
                                   df['left'].values, \
                                   test_size=0.3, \
                                   random_state=1,)
```

7. Scale the data using a **MinMaxScaler** library by running the following code:

```
from sklearn.preprocessing import MinMaxScaler

scaler = MinMaxScaler()
X_train_scaled = scaler.fit_transform(X_train)
X_test_scaled = scaler.transform(X_test)
```

When training models earlier in the notebook, we used a **StandardScaler** library, which scales each feature so that their variance is the same and they are centered around 0.

The **MinMaxScaler** library scales each feature to be in a range of 0 to 1. We can observe this by running the following code:

```
(X_train_scaled.flatten().mean(),
 X_train_scaled.flatten().min(),
 X_train_scaled.flatten().max())
```

This will print the following output:

```
(0.4231614367047082, 0.0, 1.0)
```

8. Train an SVM with the **rbf** kernel by running the following code:

```
from sklearn.svm import SVC

svm = SVC(kernel='rbf', C=1, random_state=1, gamma='scale')
svm.fit(X_train_scaled, y_train)
```

9. Calculate the classification accuracy on the test set by running the following code:

```
from sklearn.metrics import accuracy_score

y_pred = svm.predict(X_test_scaled)
accuracy_score(y_test, y_pred) * 100
```

This will print the following output, indicating an overall accuracy of ~88.8%:

```
88.84444444444445
```

10. Calculate the class accuracies on the test set by running the following code:

```
from sklearn.metrics import confusion_matrix

cmat = confusion_matrix(y_test, y_pred)
cmat.diagonal() / cmat.sum(axis=1) * 100
```

This will print the following output, indicating an accuracy of ~96% for class 0 and ~66% accuracy for class 1:

```
array([95.98946136, 66.32841328])
```

11. Plot the decision regions for the model by running the following code:

```
from mlxtend.plotting import plot_decision_regions

N_samples = 200
X, y = X_train_scaled[:N_samples], y_train[:N_samples]
plot_decision_regions(X, y, clf=svm)
plt.xlim(-0.2, 1.2)
plt.ylim(-0.2, 1.2)
```

Here's the output showing the decision regions:

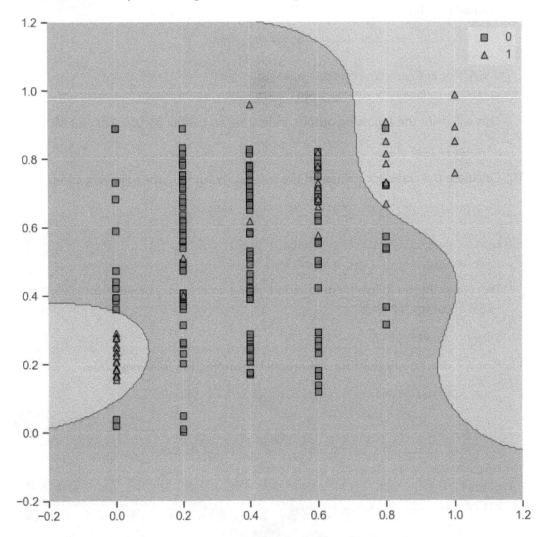

Figure 4.21: Decision region plot for a kernel SVM with C=1

12. Train an SVM with **C=50** and plot the resulting decision regions by running the following code:

```
svm = SVC(kernel='rbf', C=50, random_state=1, gamma='scale')
svm.fit(X_train_scaled, y_train)
X, y = X_train_scaled[:N_samples], y_train[:N_samples]
plot_decision_regions(X, y, clf=svm)
plt.xlim(-0.2, 1.2)
plt.ylim(-0.2, 1.2)
```

Here's the plot of an SVM with **C=50**:

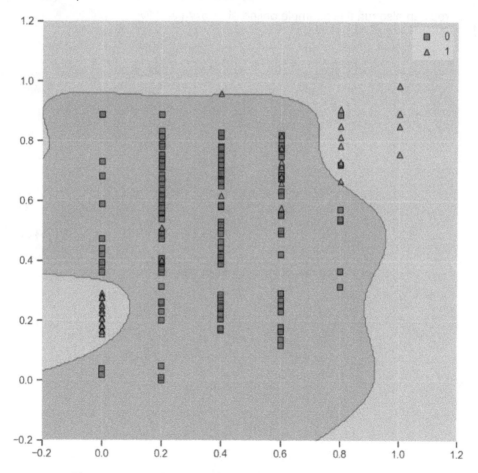

Figure 4.22: Decision region plot for a kernel SVM with C=50

Comparing the two decision region charts, we can see how the **C=50** SVM attempts to fit the patterns in the training data more closely. This is notable with respect to the point at **x, y = (0.4, 1)**, as denoted by an orange triangle. The decision surface has been adjusted to properly classify this record in the training set for the **C=50** SVM, but not for the **C=1** SVM.

> **NOTE**
>
> To access the source code for this specific section, please refer to https://packt.live/3e6JYPJ.
>
> You can also run this example online at https://packt.live/2ACdbUc.

CHAPTER 5: MODEL VALIDATION AND OPTIMIZATION

ACTIVITY 5.01: HYPERPARAMETER TUNING AND MODEL SELECTION

Solution:

1. Create a new Jupyter notebook and load the following libraries:

```
import pandas as pd
import numpy as np
import datetime
import time
import os

import matplotlib.pyplot as plt
%matplotlib inline
import seaborn as sns

%config InlineBackend.figure_format='retina'
sns.set() # Revert to matplotlib defaults
plt.rcParams['figure.figsize'] = (9, 6)
plt.rcParams['axes.labelpad'] = 10
sns.set_style("darkgrid")

%load_ext watermark
%watermark -d -v -m -p \
numpy,pandas,matplotlib,seaborn,sklearn
```

2. Load the preprocessed Human Resource Analytics dataset by running the following code:

```
df = pd.read_csv('../data/hr-analytics/hr_data_processed_pca.csv')
df.columns
```

This displays the following output:

```
Index(['satisfaction_level', 'last_evaluation', 'number_project',
       'average_montly_hours', 'time_spend_company', 'work_accident', 'left',
       'promotion_last_5years', 'department_IT', 'department_RandD',
       'department_accounting', 'department_hr', 'department_management',
       'department_marketing', 'department_product_mng', 'department_sales',
       'department_support', 'department_technical', 'salary_high',
       'salary_low', 'salary_medium', 'first_principle_component',
       'second_principle_component', 'third_principle_component'],
      dtype='object')
```

Figure 5.10: The columns of hr_data_processed_pca.csv

3. Select the features to include in the model and perform a train-test split on the data by running the following code:

```
from sklearn.model_selection import train_test_split

features = ['satisfaction_level', 'last_evaluation', \
            'time_spend_company', 'number_project', \
            'average_montly_hours', 'first_principle_component', \
            'second_principle_component', \
            'third_principle_component',]
X, X_test, \
y, y_test = train_test_split(df[features].values, \
                             df['left'].values, \
                             test_size=0.15, \
                             random_state=1)
```

4. Calculate a validation curve for the **RandomForestClassifier** with **n_estimators=50**, over the range 2 up to 52, by running the following code:

```
from sklearn.ensemble import RandomForestClassifier

np.random.seed(1)
clf = RandomForestClassifier(n_estimators=50)
max_depth_range = np.arange(2, 52, 2)
print('Training {} models ...'.format(len(max_depth_range)))
train_scores, test_scores = \
validation_curve(estimator=clf, X=X, y=y, param_name='max_depth', \
                 param_range=max_depth_range, cv=5,);
```

5. Plot the validation curve by running the following code:

```
plot_validation_curve(train_scores, test_scores, \
                      max_depth_range, xlabel='max_depth',)
plt.ylim(0.97, 1.0)
```

Here's the output of this plot:

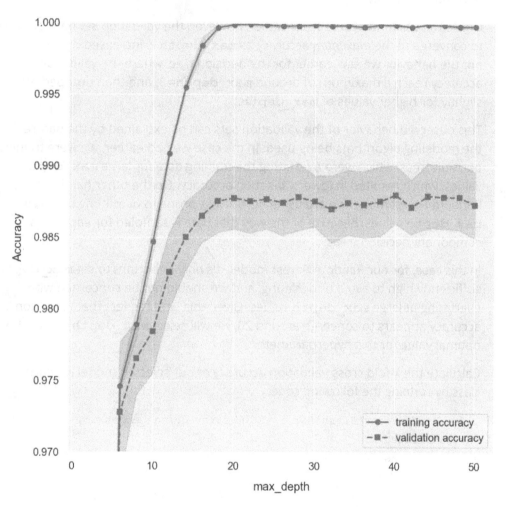

Figure 5.11: Validation curve for a Random Forest with PCA features

Here, we have an interesting result.

In some ways, this validation curve is very similar to the chart we saw for the decision tree earlier, which was trained on the same features as this Random Forest. In particular, we can see the training set (blue circles) quickly approach an accuracy of 100%, while the validation set (red square) is limited to a lower maximum accuracy.

Unlike the validation curve from earlier, however, the validation set here appears to converge to the *maximum* accuracy as **max_depth** is increased. This is not the behavior we saw earlier for the decision tree, where the validation set accuracy reach a maximum of around **max_depth=8**, and then dropped off slightly for higher values of **max_depth**.

This observed behavior of the validation sets can be explained by the nature of the modeling algorithms being used. In the case we had earlier, we were training decision trees, which were overfitting the training data for large **max_depth** values, which resulted in lower validation accuracy. On the other hand, the Random Forest model we are using here is less prone to overfitting for large **max_depth** values, because of the way that data is sampled for each of its component decision trees.

In this case, for our Random Forest model, it's only important to set **max_depth** sufficiently high to avoid underfitting, and we should not be concerned with overfitting at large **max_depth** values. Given this, and the fact that validation accuracy appears to converge around 20, we will select **max_depth=25** as the optimal value for this hyperparameter.

6. Calculate the k-fold cross validation accuracy of our selected model for each class by running the following code:

```
clf = RandomForestClassifier(n_estimators=50, max_depth=25)
np.random.seed(1)
scores = cross_val_class_score(clf, X, y)

print('accuracy = {} +/- {}'.format(scores.mean(axis=0), \
                                     scores.std(axis=0),))
```

This will print the following output (or similar, depending on how your random seeds have been set at each step):

```
fold: 1 accuracy: [0.99897119 0.94407895]
fold: 2 accuracy: [0.99897119 0.96369637]
fold: 3 accuracy: [0.99897119 0.96039604]
fold: 4 accuracy: [0.99794239 0.98349835]
fold: 5 accuracy: [0.99897119 0.96039604]
fold: 6 accuracy: [0.99691358 0.95379538]
fold: 7 accuracy: [0.99794239 0.95709571]
fold: 8 accuracy: [0.99897119 0.95709571]
fold: 9 accuracy: [0.9969104  0.94059406]
fold: 10 accuracy: [0.99485067 0.97689769]
accuracy = [0.99794154 0.95975443] +/- [0.00130286 0.01239563]
```

Comparing this to the decision tree result from the previous exercise, we can see a significant improvement in the accuracy of each class, with class 0 rising from 99.4% to 99.8% and class 1 rising from 92.4% to 95.9%.

7. Evaluate the performance of this model on the test set by running the following code:

```
from sklearn.metrics import confusion_matrix

clf = RandomForestClassifier(n_estimators=50, max_depth=25)
clf.fit(X, y)

y_pred = clf.predict(X_test)
cmat = confusion_matrix(y_test, y_pred)
cmat.diagonal() / cmat.sum(axis=1) * 100
```

This will print the following output:

```
array([99.70760234, 97.03703704])
```

By comparing these numbers with the k-fold accuracies from before, we can see that they both lie in the expected range, hence verifying this model.

8. Train the model on the full set of records in **df** by running the following code:

```
features = ['satisfaction_level', 'last_evaluation', \
            'time_spend_company', 'number_project', \
            'average_montly_hours', 'first_principle_component', \
            'second_principle_component', \
            'third_principle_component',]
X = df[features].values
y = df['left'].values


clf = RandomForestClassifier(n_estimators=50, max_depth=25)
clf.fit(X, y)
```

9. Save the model to disk by running the following code:

```
import joblib
joblib.dump(clf, 'hr-analytics-pca-forest.pkl')
```

Then, reload the model, as follows:

```
clf = joblib.load('hr-analytics-pca-forest.pkl')
clf
```

This should return the string representation of the trained model:

```
RandomForestClassifier(bootstrap=True, class_weight=None, criterion='gini',
                       max_depth=25, max_features='auto', max_leaf_nodes=None,
                       min_impurity_decrease=0.0, min_impurity_split=None,
                       min_samples_leaf=1, min_samples_split=2,
                       min_weight_fraction_leaf=0.0, n_estimators=50,
                       n_jobs=None, oob_score=False, random_state=None,
                       verbose=0, warm_start=False)
```

Figure 5.12: Random Forest model representation

10. Check the model's performance for an imaginary employee, Alice, by selecting the appropriate features from row 573 of **df** with the following code:

```
alice = df.iloc[573][features]
alice
```

This will print the following output, representing the employee metrics and derived principal components for Alice:

```
satisfaction_level            0.360000
last_evaluation               0.470000
time_spend_company            3.000000
number_project                2.000000
average_montly_hours        148.000000
first_principle_component     0.742801
second_principle_component   -0.514568
third_principle_component    -0.677421
```

11. Predict the class label for Alice, as follows:

```
clf.predict([alice.values])
```

This will print the following output:

```
array([1])
```

Then, calculate the probability assigned to this prediction, as follows:

```
clf.predict_proba([alice.values])
```

This will print the following output:

```
array([[0., 1.]])
```

These results indicate that the model predicts with 100% probability that Alice will leave the company.

12. In order to improve the chance of the company being able to retain Alice as an employee, they could try to reduce the amount of time she needs to spend at work. Using our model, we can test the effect that might have on her likelihood of leaving.

Set **average_montly_hours=100** and **time_spend_company=2** and then re-evaluate the prediction probability of the model by running the following code:

```
alice.average_montly_hours = 100
alice.time_spend_company = 2
clf.predict_proba([alice.values])
```

This will print the following output:

```
array([[0.84, 0.16]])
```

Predict the new class label with the following command:

```
clf.predict([alice.values])
```

This will print the following output:

```
array([0])
```

This result suggests that by reducing the number of monthly hours at work to 100 and the amount of time spent at the company to level 2, there's an 84% chance that Alice will not leave the company.

> **NOTE**
>
> To access the source code for this specific section, please refer to https://packt.live/37vgad6.
>
> You can also run this example online at https://packt.live/2BcG5tP.

This is a great example of how predictive modeling can be used by businesses to make data-driven decisions.

CHAPTER 6: WEB SCRAPING WITH JUPYTER NOTEBOOKS

ACTIVITY 6.01: WEB SCRAPING WITH JUPYTER NOTEBOOK

Solution:

1. Run the following code in your notebook to load the necessary libraries:

```
import pandas as pd
import numpy as np
import datetime
import time
import os

import matplotlib.pyplot as plt
%matplotlib inline
import seaborn as sns
import requests
from bs4 import BeautifulSoup

%config InlineBackend.figure_format='retina'
sns.set() # Revert to matplotlib defaults
plt.rcParams['figure.figsize'] = (9, 6)
plt.rcParams['axes.labelpad'] = 10
sns.set_style("darkgrid")

%load_ext watermark
%watermark -d -v -m -p \
requests,numpy,pandas,matplotlib,seaborn,sklearn
```

2. After defining the **url** variable, load that page in the notebook using an IFrame. This can be done by running the following code:

```
url = 'https://en.wikipedia.org/wiki/List_of_countries_and'\
      '_dependencies_by_population'
from IPython.display import IFrame
IFrame(url, height=300, width=800)
```

Here's the output displaying IFrame:

WIKIPEDIA
The Free Encyclopedia

Article Talk

Read Edit View history   Search W

List of countries and dependencies by population

From Wikipedia, the free encyclopedia

Main page
Contents
Featured content
Current events
Random article
Donate to Wikipedia

This is a **list of countries and dependent territories by population**. It includes sovereign states, inhabited dependent territories and, in some cases, constituent countries of sovereign

Figure 6.29: Running the live web page in Jupyter

When you have finished browsing the page, close it by clicking into the cell and selecting **Cell | Current Outputs | Clear**.

3. Request the page by running the following code:

```
resp = requests.get(url)
resp
```

This should print the following output:

```
<Response [200]>
```

4. Instantiate a **BeautifulSoup** object using the response content by running the following command:

```
soup = BeautifulSoup(resp.content, 'html.parser')
```

5. Get the **h1** of the page by running the following command:

```
soup.find_all('h1')
```

This should print the following output:

```
[<h1 class="firstHeading" id="firstHeading" lang="en">List of countries
and dependencies by population</h1>]
```

6. Select the **div** with **id="bodyContent"** by running the following command:

```
body_content = soup.find('div', {'id': 'bodyContent'})
```

7. Search this div element for the table headers by running the following code:

```
table_headers = body_content.find_all('th')
table_headers
```

Here's the output showing display headers:

```
[<th data-sort-type="number">Rank</th>,
 <th>Country (or dependent territory)</th>,
 <th>Population</th>,
 <th>% of world<br/>population</th>,
 <th>Date</th>,
 <th class="unsortable">Source
 </th>,
 <th>1
 </th>,
 <th>2
 </th>,
 <th>3
 </th>,
 <th>4
 </th>,
 <th>5
 </th>,
 <th>6
 </th>,
```

Figure 6.30: The table header elements

Then, print the element text for each, along with their index, by running the following code:

```
for i, t in enumerate(table_headers):
    print(i, t.text.strip())
    print('-'*10)
```

This loop is displayed as follows:

```
0 Rank
- - - - - - - - - -
1 Country (or dependent territory)
- - - - - - - - - -
2 Population
- - - - - - - - - -
3 % of worldpopulation
- - - - - - - - - -
4 Date
- - - - - - - - - -
5 Source
- - - - - - - - - -
6 1
- - - - - - - - - -
7 2
- - - - - - - - - -
8 3
- - - - - - - - - -
9 4
- - - - - - - - - -
```

Figure 6.31: Printing the text of each table header element

8. Manually set the table headers by running the following code:

```
table_headers = ['Country(or dependent territory)', 'Population', \
                 '% of WorldPopulation', 'Date']
```

9. Select the data entries from the row element at index **2** by running the following code:

```
row_number = 2
row_data = body_content.find_all('tr')[row_number]\
           .find_all(['td', 'th'])
```

10. Find the number of data entries in the row by running **len(row_data)**. The result is **6**.

11. Print out the row's data entries by running **row_data**:

```
[<th>2
</th>,
<td align="left"><span class="flagicon"><img alt="" class="thumbborder" data-file-height="900" data-
file-width="1350" decoding="async" height="15" src="//upload.wikimedia.org/wikipedia/en/thumb/4/41/Fl
ag_of_India.svg/23px-Flag_of_India.svg.png" srcset="//upload.wikimedia.org/wikipedia/en/thumb/4/41/Fl
ag_of_India.svg/35px-Flag_of_India.svg.png 1.5x, //upload.wikimedia.org/wikipedia/en/thumb/4/41/Flag_
of_India.svg/45px-Flag_of_India.svg.png 2x" width="23"/></span> <a href="/wiki/Demographics_of_India"
title="Demographics of India">India</a><sup class="reference" id="cite_ref-6"><a href="#cite_note-6">
[c]</a></sup></td>,
<td style="text-align:right">1,364,925,239</td>,
<td align="right"><span data-sort-value="7001174981800767757♠" style="display:none"></span>17.5%</td
>,
<td><span data-sort-value="000000002020-07-20-0000" style="white-space:nowrap">20 Jul 2020</span></t
d>,
<td align="left">National population clock<sup class="reference" id="cite_ref-7"><a href="#cite_note
-7">[4]</a></sup>
</td>]
```

Figure 6.32: The data elements in a selected row

12. Print the element text for each row data entry, along with their index, by running the following code:

```
for i, row in enumerate(row_data):
    print(i, row.text)
```

The output is seen in the following screenshot:

```
0 2

1  India[c]
2 1,364,925,239
3 17.5%
4 20 Jul 2020
5 National population clock[4]
```

Figure 6.33: Printing the text for each data element in a selected row

13. Select the fields we are interested in parsing by running the following code:

```
row_data = row_data[1:5]
```

14. First, assign **d1** to the first element and parse the entry value using the following code:

```
d1 = row_data[0]
d1.find('a').text
```

Here, we used the trick of searching for an **<a>** element in the data entry, before getting the text.

15. Next, do the same for the second entry:

```
d2 = row_data[1]
d2.text
```

16. Then, do the same for the third entry:

```
d3 = row_data[2]
d3.text
```

17. Finally, do the same for the last entry:

```
d4 = row_data[3]
d4.text
```

For the last three entries, we are happy with simply getting the element text for now. They can be processed further (for example, by converting strings into numeric values) at a later time.

18. Perform the full scrape of country populations by running the following code:

```
# Perform the full scrape by iterating over the rows

pop_data = []
row_elements = body_content.find_all('tr')
for i, row in enumerate(row_elements):
    row_data = row.find_all(['td','th'])
    if len(row_data) < 5:
        print('Ignoring row {} because length < 5'.format(i))
        continue
```

```python
d1, d2, d3, d4 = row_data[1:5]
errs = []

try:
    d1 = d1.find('a').text
except Exception as e:
    d1 = ''
    errs.append(str(e))

try:
    d2 = d2.text
except Exception as e:
    d2 = ''
    errs.append(str(e))

try:
    d3 = d3.text
except Exception as e:
    d3 = ''
    errs.append(str(e))

try:
    d4 = d4.text
except Exception as e:
    d4 = ''
    errs.append(str(e))
data = [d1, d2, d3, d4]
print(data)
pop_data.append(data)

if errs:
    print('Errors in row {}: {}'.format(i, ', '.join(errs)))
```

19. Print out the data we parsed previously, by running **pop_data**. The output is as follows:

```
[['', 'Population', '% of worldpopulation', 'Date'],
 ['China', '1,403,612,840', '18.0%', '20 Jul 2020'],
 ['India', '1,364,925,239', '17.5%', '20 Jul 2020'],
 ['United States', '329,985,664', '4.23%', '20 Jul 2020'],
 ['Indonesia', '269,603,400', '3.46%', '1 Jul 2020'],
 ['Pakistan', '220,892,331', '2.83%', '1 Jul 2020'],
 ['Brazil', '211,817,730', '2.72%', '20 Jul 2020'],
 ['Nigeria', '206,139,587', '2.64%', '1 Jul 2020'],
 ['Bangladesh', '168,985,154', '2.17%', '20 Jul 2020'],
 ['Russia', '146,748,590', '1.88%', '1 Jan 2020'],
 ['Mexico', '127,792,286', '1.64%', '1 Jul 2020'],
 ['Japan', '125,930,000', '1.61%', '1 Jun 2020'],
 ['Philippines', '108,916,944', '1.40%', '20 Jul 2020'],
 ['Egypt', '100,640,472', '1.29%', '20 Jul 2020'],
 ['Ethiopia', '98,665,000', '1.26%', '1 Jul 2019'],
```

Figure 6.34: Output of the pop_data command

20. Print out the table headers by running **table_headers**. The output is as follows:

```
['Country(or dependent territory)',
 'Population',
 '% of WorldPopulation',
 'Date']
```

21. Save the data in a CSV file by running the following code:

```
f_path = '../data/countries/populations_raw.csv'
pd.DataFrame(pop_data, columns=table_headers)\
            .to_csv(f_path, index=False)
```

22. If your Jupyter environment supports bash, then you can view the head of your table by running the following code:

```
%%bash
head ../data/countries/populations_raw.csv
```

The **head** is displayed as follows:

```
Country(or dependent territory),Population,% of WorldPopulation,Date
,Population,% of worldpopulation,Date
China,"1,403,612,840",18.0%,20 Jul 2020
India,"1,364,925,239",17.5%,20 Jul 2020
United States,"329,985,664",4.23%,20 Jul 2020
Indonesia,"269,603,400",3.46%,1 Jul 2020
Pakistan,"220,892,331",2.83%,1 Jul 2020
Brazil,"211,817,730",2.72%,20 Jul 2020
Nigeria,"206,139,587",2.64%,1 Jul 2020
Bangladesh,"168,985,154",2.17%,20 Jul 2020
```

Figure 6.35: The first 10 rows of extracted Wikipedia data in a CSV file

> **NOTE**
>
> To access the source code for this specific section, please refer to
> https://packt.live/2ACHg63.
>
> You can also run this example online at https://packt.live/2zDrqYu.

To summarize, we've seen how Jupyter Notebooks can be used for web scraping. We started this chapter by learning about HTTP methods and status codes. Then, we used the **requests** library to actually perform HTTP requests with Python and saw how the **BeautifulSoup** library can be used to parse response HTML.

ACTIVITY 6.02: ANALYZING COUNTRY POPULATIONS AND INTEREST RATES

Solution

1. Start up one of the following platforms for running Jupyter Notebooks:

   JupyterLab (run **jupyter lab**)

   Jupyter Notebook (run **jupyter notebook**)

   Then, open up the chosen platform in your web browser by copying and pasting the URL, as prompted in the Terminal.

2. Run the following code to load some of the libraries that we'll be using to configure our plot settings for the notebook:

```
import pandas as pd
import numpy as np
import datetime
import time
import os

import matplotlib.pyplot as plt
%matplotlib inline
import seaborn as sns

%config InlineBackend.figure_format='retina'
sns.set() # Revert to matplotlib defaults
plt.rcParams['figure.figsize'] = (9, 6)
plt.rcParams['axes.labelpad'] = 10
sns.set_style("darkgrid")

%load_ext watermark
%watermark -d -v -m -p numpy,pandas,matplotlib,seaborn
```

3. Load the processed country population and interest rate dataset by running the following code:

```
df = pd.read_csv('../data/countries/country_data_merged.csv')
df.head()
```

The output displaying the head is shown in the following screenshot:

	country	population	population_pct	date_population_update	interest_rate_pct	date_interest_rate_last_change
0	China	1.403613e+09	18.00	2020-07-20	4.20	2019-09-20
1	United States	3.299857e+08	4.23	2020-07-20	0.25	2020-05-06
2	Indonesia	2.696034e+08	3.46	2020-07-01	4.50	2020-03-19
3	Pakistan	2.208923e+08	2.83	2020-07-01	9.00	2020-04-17
4	Brazil	2.118177e+08	2.72	2020-07-20	2.25	2020-06-17

Figure 6.36: The head of country_data_merged.csv

4. Print the datatypes of each column by running **df.dtypes**:

```
country                         object
population                      float64
population_pct                  float64
date_population_update          object
interest_rate_pct               float64
date_interest_rate_last_change  object
average_inflation_rate_pct      float64
dtype: object
```

Figure 6.37: The datatypes of each column

5. Convert the date column entries into datetimes by running the following code:

```
df['date_population_update'] = \
pd.to_datetime(df['date_population_update'])


df['date_interest_rate_last_change'] = \
pd.to_datetime(df['date_interest_rate_last_change'])
```

6. Check for missing values by running the following command:

```
df.isnull().sum()
```

The following screenshot shows the output of this command:

```
country                         0
population                      0
population_pct                  0
date_population_update          0
interest_rate_pct               0
date_interest_rate_last_change  0
average_inflation_rate_pct      0
dtype: int64
```

Figure 6.38: The number of missing values in each column

7. Plot comparable histograms of the date columns, that is, **date_population_update** and **date_interest_rate_last_change**, by running the following code:

```
col = 'date_population_update'
df[col].hist(bins=10, alpha=0.7, label=col)


col = 'date_interest_rate_last_change'
df[col].hist(bins=45, alpha=0.7, label=col)


plt.xticks(rotation=45)
plt.legend()
plt.show()
```

The histogram plotted from this code is seen here:

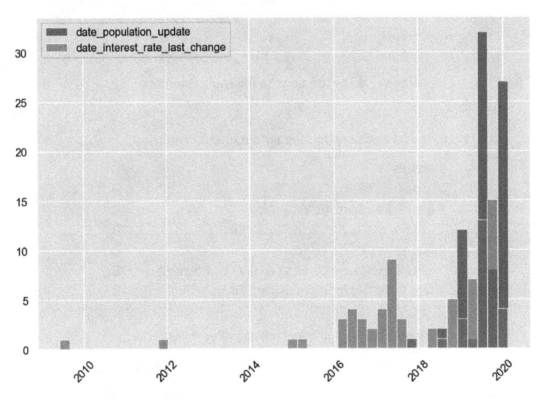

Figure 6.39: Histograms for the date fields in the dataset

8. Find the country with the oldest date since the last interest rate change by running the following code:

```
min_date = df['date_interest_rate_last_change'].min()
min_date_mask = df['date_interest_rate_last_change'] == min_date
df[min_date_mask]
```

Here's the output of this code:

	country	population	population_pct	date_population_update	interest_rate_pct	date_interest_rate_last_change
80	Barbados	287025.0	0.00368	2019-07-01	7.0	2009-06-01

Figure 6.40: Country wth the oldest date since last interest rate change

> **NOTE**
>
> The preceding DataFrame is cropped for representation purpose. You can refer to the complete DataFrame at https://packt.live/2ACHg63.

9. Find any countries that have negative interest rates by running the following code:

```
neg_rates_mask = df['interest_rate_pct'] < 0
df[neg_rates_mask][['country', 'interest_rate_pct']]
```

The countries with negative interest rates are shown here:

	country	interest_rate_pct
9	Japan	-0.10
51	Switzerland	-0.75
58	Denmark	-0.75

Figure 6.41: Displaying the countries with negative interest rates

10. Plot a bar chart of the top countries by population, by running the following code:

```
df_plot = df.sort_values('population', ascending=False).head(10)

df_plot['population'].plot.bar()

plt.xticks(rotation=45)
plt.ylabel('Population')

ax = plt.gca()
labels = df_plot['country'].values
ax.set_xticklabels(labels)
plt.show()
```

The bar chart thus plotted is shown as follows:

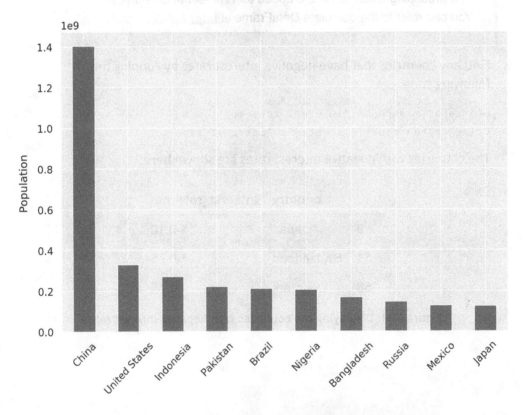

Figure 6.42: A bar chart comparing the top countries by population

11. Plot a scatter chart of **population_pct** versus **interest_rate_pct** by running the following code:

```
sns.scatterplot(data=df, x='population_pct', y='interest_rate_pct')
plt.show()
```

The scatter chart plotted using this code is displayed here:

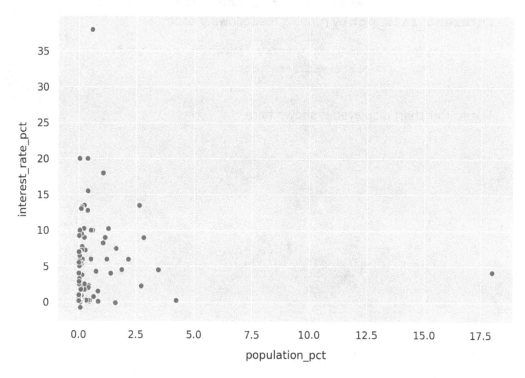

Figure 6.43: A scatter chart showing interest rates as a function of population

12. Select the rows that correspond to the outlier points along the population axis shown in the preceding chart by running the following command:

```
df[df['population_pct'] > 15]
```

Here's the output showing outliers:

	country	population	population_pct	date_population_update	interest_rate_pct	date_interest_rate_last_change	average_inflation_rate_pct
0	China	1.403555e+09	18.0	2020-07-16	4.2	2019-09-20	1.93

Figure 6.44: The countries that have over 15% of the world population

> **NOTE**
>
> The preceding DataFrame is cropped for representation purpose.
> You can refer to the complete DataFrame at https://packt.live/2ACHg63.

13. Plot a scatter chart of **average_inflation_rate_pct** versus **interest_rate_pct** by running the following code:

```
sns.scatterplot(data=df, x='average_inflation_rate_pct', \
                y='interest_rate_pct')
plt.show()
```

The scatter chart displayed is shown here:

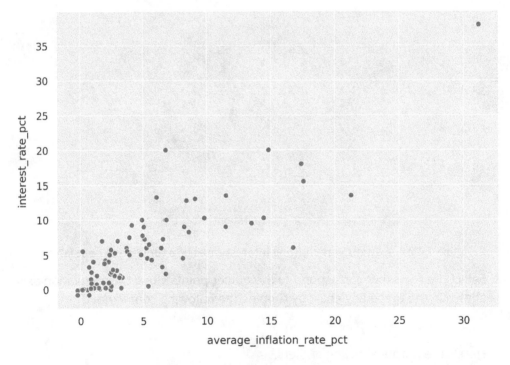

Figure 6.45: A scatter chart showing interest rates as a function of average inflation

That concludes this activity. However, we can take this further to demonstrate the kind of modeling that can be carried out once data has been loaded, explored, and visualized. Let's see how the preceding data can be used to fit a k-means clustering model. k-means clustering has not been covered in any of these chapters, as it is out of scope of this book; however, you are encouraged to follow the code and use it as a basis for further learning and exploration. Some of the tools and techniques that we will use here will be unfamiliar, but you can carry out your own research to develop your understanding. The steps are as follows:

14. Use scikit-learn's **StandardScaler** class to prepare the features for modeling by running the following code:

```
from sklearn.preprocessing import StandardScaler
scaler = StandardScaler()


features = ['population_pct', 'interest_rate_pct', \
            'average_inflation_rate_pct']
X = df[features].values
X_scaled = scaler.fit_transform(X)


df['scaled_population_pct'] = X_scaled[:,0]
df['scaled_interest_rate_pct'] = X_scaled[:,1]
df['scaled_average_inflation_rate_pct'] = X_scaled[:,2]
```

After importing and instantiating the scaler class, we select our feature values from the DataFrame and feed them into the **fit_transform** method. By doing this, the scaler learns what the data looks like and determines how to scale it properly, according to the underlying algorithm. It returns the scaled version of our features, which are then mapped back onto the DataFrame.

15. Now, let's train a clustering model on the **population_pct, interest_rate_pct**, and **average_inflation_rate_pct** features. We'll start with the following code:

```
from sklearn.cluster import KMeans
clf = KMeans(n_clusters=5)
clf.fit(X_scaled)
```

16. Label each row of the training data by running the following command:

```
df['kmeans_cluster'] = clf.predict(X_scaled)
```

17. Plot a scatter chart of **population_pct** versus **interest_rate_pct**, where colors are assigned according to clusters, by running the following code:

```
sns.scatterplot(data=df, x='population_pct', \
                y='interest_rate_pct', hue='kmeans_cluster',)
plt.show()
```

This will display the following scatter chart:

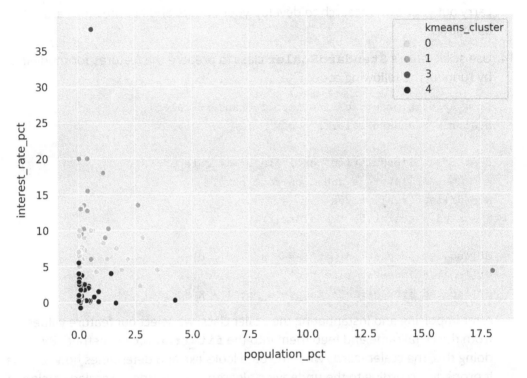

Figure 6.46: Segmenting the interest rate versus population chart with KMeans clusters

18. Plot a scatter chart of **average_inflation_rate_pct** versus **interest_rate_pct**, where colors are assigned according to clusters, by running the following code:

```
sns.scatterplot(data=df, x='average_inflation_rate_pct', \
                y='interest_rate_pct', hue='kmeans_cluster',)
plt.show()
```

This will display the following scatter plot:

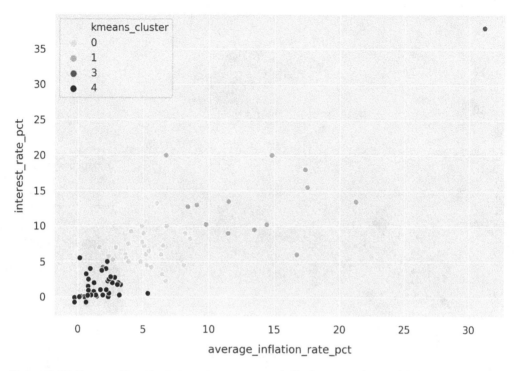

Figure 6.47: Segmenting the interest rate versus inflation rate chart with KMeans clusters

19. For the previous chart, add point sizes corresponding to country populations by running the following code:

```
sns.scatterplot(data=df, x='average_inflation_rate_pct', \
                y='interest_rate_pct', size='population_pct', \
                hue='kmeans_cluster', sizes=(20, 500),)
plt.xlim(-1, 15)
plt.ylim(-1, 15)
plt.show()
```

The scatter chart produced is shown here:

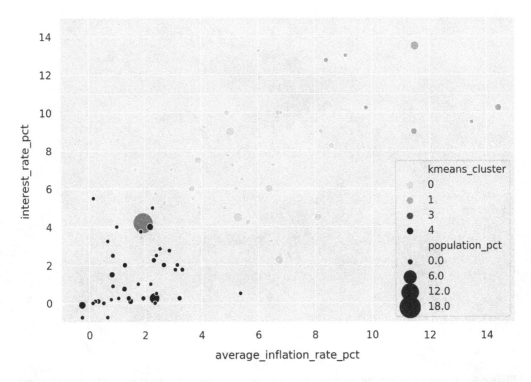

Figure 6.48: Overlaying population data of the KMeans segmented chart of interest rates versus inflation rates

20. Create a three-dimensional version of the preceding chart by plotting the scatter chart of **average_inflation_rate_pct** versus **interest_rate_pct** versus **population_pct** by defining the following function:

```
def render_3d_plot():
    from mpl_toolkits.mplot3d import Axes3D
    import matplotlib.pyplot as plt
    plt.rcParams['figure.figsize'] = (9, 9)
    fig = plt.figure()

    ax = fig.add_subplot(111, projection='3d')
    ax.scatter(df['interest_rate_pct'], \
               df['average_inflation_rate_pct'], \
               df['population_pct'], \
               c=df['kmeans_cluster'].values,)
    ax.set_xlabel('interest_rate_pct')
    ax.set_ylabel('average_inflation_rate_pct')
    ax.set_zlabel('population_pct')
```

When using the Jupyter Notebook platform, a static version of the plot can be rendered by running the following code:

```
%matplotlib inline
render_3d_plot()
```

The output of this code is shown here:

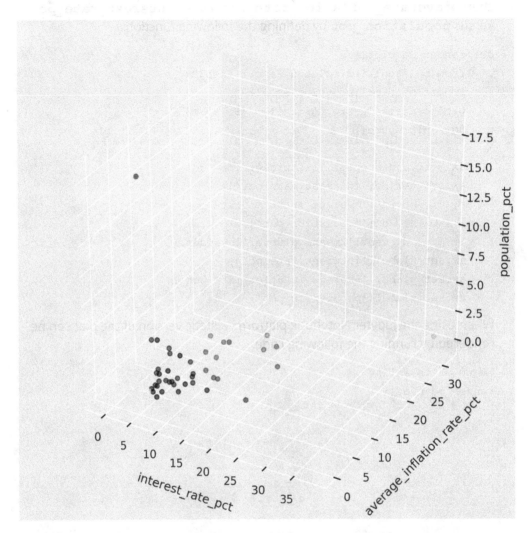

Figure 6.49: Visualizing the KMeans clustering model with a three-dimensional scatter chart

21. When using the Jupyter Notebook platform, an interactive version of the plot can be rendered by running the following code:

```
%matplotlib notebook
render_3d_plot()
```

In order to get this working in JupyterLab, you may need to install the **jupyter-matplolib** extension. You can find the installation instructions on GitHub at https://packt.live/2UNrzzQ. Once installed, you can run the following code in JupyterLab to render the interactive chart:

```
%matplotlib widget
render_3d_plot()
```

The output is as follows:

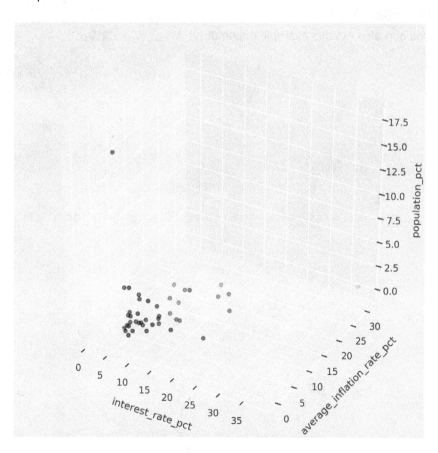

Figure 6.50: Visualizing the k-Means clustering model with an interactive three-dimensional scatter chart

22. Save the model using **`joblib`** by running the following code:

```
import joblib
joblib.dump(scaler, 'kmeans-5-cluster-scaler.pkl')
joblib.dump(clf, 'kmeans-5-cluster-model.pkl')
```

In order to use this model for making classifications, we need to save both the trained model, **clf**, and the trained **scaler**. This way, future data can be scaled properly before we feed it into **clf**. This final step concludes the bonus material and full activity solution.

> **NOTE**
>
> To access the source code for this specific section, please refer to https://packt.live/2ACHg63.
>
> You can also run this example online at https://packt.live/2zDrqYu.

INDEX

www.ingramcontent.com/pod-product-compliance
Lightning Source LLC
Chambersburg PA
CBHW062055050326
40690CB00016B/3102